THE
ELECTRIC LIFE
OF
MICHAEL FARADAY

Faraday, from an 1840s daguerreotype by Antoine Claudet.

THE
ELECTRIC LIFE
OF
MICHAEL FARADAY

Alan Hirshfeld

WALKER & COMPANY

New York

First published in the United States of America in 2006 by
Walker Publishing Company, Inc.
Distributed to the trade by Holtzbrinck Publishers.

For information about permission to reproduce selections from
this book, write to Permissions, Walker & Company,
104 Fifth Avenue, New York, New York 10011.

All papers used by Walker & Company Publishing are natural, recyclable
products made from wood grown in well-managed forests.
The manufacturing processes conform to the environmental
regulations of the country of origin.

Library of Congress Cataloging-in-Publication Data

Hirshfeld, Alan.
The electric life of Michael Faraday / Alan Hirshfeld.
    p.   cm.
Includes bibliographical references and index.
ISBN-13: 978-0-8027-1470-1
ISBN-10: 0-8027-1470-6
1. Faraday, Michael, 1791–1867.   2. Physicists—Great Britain—
Biography.   I. Title.
QC16.F2H57   2006
530.092—dc22
[B]
2005025533

Visit Walker & Company's Web site at www.walkerbooks.com

Typeset by Westchester Book Group

Printed in the United States of America by Quebecor World Fairfield

2   4   6   8   10   9   7   5   3   1

For Sally, who taught me to riff

CONTENTS

Science is not a collection of facts, any more than opera is a collection of notes. It's a process, a way of thinking, a method based on a single insight—that the degree to which an idea seems true has nothing to do with whether it is true, and that the way to distinguish factual ideas from false ones is to test them by experiment.

—TIMOTHY FERRIS, "NOT ROCKET SCIENCE," *The New Yorker* (JULY 20, 1998), P. 5

[E]ven if I could be Shakespeare, I think I should still choose to be Faraday.

—ALDOUS HUXLEY

PREFACE

The capsule version of Michael Faraday's life reads like a fairy tale: Through sheer gumption and timely luck, a poor, unschooled bookbinder's apprentice in nineteenth-century England surmounted adversity and class prejudice to become the greatest experimental scientist of his time. We are the beneficiaries of Faraday's good fortune. The electric motors that raise our elevators, spin our fans, and propel our hybrid cars trace their origins to Faraday's laboratory. The generators that electrify our reading lights and computers stem from the copper disk Faraday spun between the poles of a magnet more than a century and a half ago. The transformers that enable electricity to race hundreds of miles across the countryside and then safely into our homes are descendents of Faraday's wire-encased iron rings. And the development of the theoretical principles underlying these electrical, magnetic, and luminous marvels was inspired by Faraday's farsighted speculations.

Faraday's particular gift as an experimentalist was to make visible in the laboratory that which had been invisible, to magnify nature's subtle effects so they could be perceived and measured. He vaulted to fame in the 1830s after developing a host of ideas, processes, and devices that undergird modern technology. Then he embarked on a decades-long quest for the holy grail of nineteenth-century physics: a comprehensive theory of electricity, magnetism, force, and light. Here he entered a realm where experimental verification was difficult, if not impossible—the realm of the mathematician, who solves equations to find plausible explanations for physical phenomena. And Faraday, facile as he was in the laboratory, was a grade-school mathematician. The marathon of experimentation and cogitation produced a raft of important scientific results—but also broke Faraday's health. He suspended his research for five gloomy years, only to return with a triumphant experiment showing that magnetism affects light. It was not until the twilight of Faraday's career that a young admirer, James Clerk Maxwell, cast his hero's controversial speculations about the nature of light and the transmission of force into the abstract symbolism of mathematics. The result—Maxwell's famous field equations—was a radically new vision of nature, whose ramifications not only inspired a new generation of physicists, including Einstein, but still resound today.

If there was one overriding element to Faraday's character, it was *humility*. His "conviction of deficiency," as he called it, stemmed in part from his deep religiosity and affected practically every facet of his life. Thus Faraday approached both his science and his everyday conduct unhampered by ego, envy, or negative emotion. In his work, he assumed the inevitability of error and failure; whenever possible, he harnessed these as guides toward further investigation. Faraday adhered to no particular school of scientific thought. Nor did he flinch when a favored hypothesis fell to the rigors of experiment. In the personal realm, Faraday subjected himself to constant self-examination and correction. Only his belief in God rested solely on unquestionable faith. Although devout, he kept a strict separation between his religious practice and the methods of science. In fact, to reveal nature's design through scientific study, in his opinion, only affirmed the glory of God. Religion provided motivation, not method, in Faraday's work.

Faraday also made contributions in civic affairs beyond his renowned science lectures for the public. Toward the end of his career, Faraday used the weight of his reputation to crusade for science education and environmental responsibility. His views on these subjects sound remarkably apt today, a century and a half later. His complaints about the pollution of the Thames River triggered long-term efforts to improve water quality. Faraday also led a quixotic charge against the public's propensity toward superstition and pseudoscience, which he denounced as a "disgrace to the age." He would be similarly appalled by today's New Age hokum and by attempts of extremists to inject religious ideology into scientific education.

My interest in Michael Faraday runs deep, not just in my capacity as a physics professor at the University of Massachusetts, Dartmouth, but in the arc of my career. Like Faraday, I felt the tug of science at a young age. I, too, immersed myself in books on science and sought out mentors to teach me what I needed to know. I, too, aspired to a life exploring—and explaining to others—the wonders of nature. The name "Faraday" has been in my consciousness since at least 1970, when I took freshman physics and studied his experiments in electricity and magnetism. Yet it was in researching this book during the past two years that I truly discovered who he was.

The more I got to know Faraday through his many scientific reports, letters, journals, and diaries, the more I enjoyed "having him around." I felt as though I were there in his laboratory, peering over his shoulder as he unraveled nature's mysteries. And at his Friday evening lectures, witnessing spectacular science demonstrations. And on his walks in the countryside, admiring the hue of the sunset or the spectacle of a thunderstorm. I shared his joy in discovery, his pain in failure, his sheer exuberance in the scientific endeavor. Frankly, at times my identification with Faraday went too far, as when I was writing the chapter about his collapse from overwork—only to suffer a similar bout of exhaustion myself. So much a living spirit did he become to me that I found it difficult at the end of the book to pen his death. (I felt compelled to "resurrect" him in an epilogue.)

In sum, Michael Faraday was one of those rare scientists in the mold of Galileo, Newton, and Einstein, free of blinding preconceptions about nature and thus endowed with a vision denied his contemporaries. Yet he was also an everyman with whom we can identify: the dreamer whose dream comes true; the genius whose genius falls short; the human whose humanity rules his actions. Faraday was truly one of us.

1

IMPROVEMENT OF
THE MIND

*Some see nature all ridicule and deformity . . . and some
scarce see nature at all. But to the eyes of the man of
imagination, nature is imagination itself.*

—WILLIAM BLAKE

Step back two centuries and through the front door of George
Riebau's bookbindery at 2 Blandford Street near Manchester
Square in London. Smell the pungent aroma of leather, glue, and
varnish. Hear the murmurous drumbeat of the binder's mallet
tamping gathered pages. Books are everywhere—on shelves, on tables, even wedged into the cubbylike window frames, where they
eclipse the light struggling to enter. In this dim paper-and-leather
universe of long ago, Riebau and his three apprentices stand at their
benches, plying the bookbinder's craft. Around them lie the accou-
trements of their trade: needles, thread, Jaconette cloth, engraving
tools, standing press, cutting boards. The room buzzes with conversation, for Riebau is a genial man who likes to keep his workers and
his customers happy. Yet for all the chatter, the binding and selling
of books appear to be the sole order of business here. In short,
George Riebau's modest establishment is the last place one would
suspect as an incubator for a would-be scientist—especially in 1812.

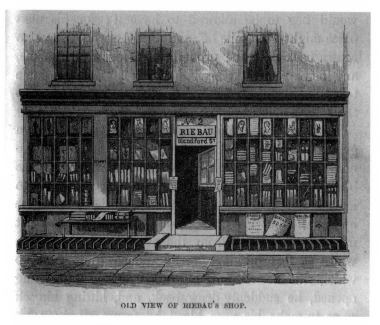

OLD VIEW OF RIEBAU'S SHOP.

The London bookbindery where Faraday apprenticed until he was twenty-one.

But move beyond the benches to the small fireplace that keeps the workers' fingers supple through the frigid London winters. There on the mantelpiece, arrayed in no particular order, is a curious assortment of devices that bear no connection to the binding of books: voltaic piles—batteries, in today's parlance; copper and zinc electrodes; coils of wire; bottled acids; glass cylinders for generating and storing electricity. Nearby, meticulous pencil sketches of electrical machines. Alongside these, jottings about electrical phenomena. Here is the after-hours "laboratory" of young Michael Faraday, one of George Riebau's apprentices, who is at present probably counting the minutes until he can set aside his tools and resume his homespun experiments. Faraday's teachers, such as they are, do not wear silken robes or roam ivy-covered buildings; they speak to Faraday silently from the printed pages that pass through his hands on the way to more advantaged customers. To Faraday, Riebau's shop is truly library, classroom, and laboratory. The mantelpiece curios are manifestations of a dream by a young man whose ambitions are pressing ever more despairingly against the harsh realities of British society. This in an age when the term "upward mobility" holds no practical meaning for the mass of humanity—when, for the most part, scientists are born, not made.

In a few short months, Michael Faraday's apprenticeship will end and he must take up the career for which he trained: bookbinding. And therein lies the source of his searing realization that his life might be spent in the mindless packaging of countless words on countless subjects, and not one of his own devising. For Faraday longs to uncover nature's secrets, not as a hobbyist in some dusty shop corner, but as a professional man of science in a real laboratory. Now with both time and hope running short, he can only wait for opportunity to extend its hand and, God willing, sweep him into the ranks of England's scientific elite.

In the winter of 1791, blacksmith James Faraday moved his wife and two children from Outhgill in northern England to what was then the village of Newington Butts, now the Elephant and Castle section of London just south of the Thames. By the time his third child, Michael, was born in September, the brighter prospects James Faraday was seeking had not yet materialized. Nor did they ever.

In fragile health, James Faraday worked intermittently, and at times only the charity of others kept the family from outright starvation. Nevertheless, the Faradays were buoyed by their bedrock religious faith and belief in the blessings of a simple, if impoverished, life. In 1796, the family, now six in number, moved to cramped quarters over a stable in Jacob's Well Mews in central London. As a child, Michael Faraday passed the hours running with the neighborhood boys, shooting marbles in nearby Spanish Place, or entertaining his baby sister Peggy in Manchester Square. Schooling was brief: Faraday's mother withdrew him when the schoolmistress attempted to cane a speech defect out of him. (He could not pronounce his *r*'s, and referred to his brother as "Wobert.") Occasionally he visited his father at James Boyd's forge a few blocks east on Welbeck Street. Once, while playing in the loft, he fell through a hole onto his father's back, averting sure disaster on the anvil's edge. In 1804, thirteen-year-old Faraday was hired as an errand boy for George Riebau, a French émigré, political radical, and proprietor of a book shop around the block. A year later, on October 7, 1805, Faraday entered a seven-year apprenticeship with Riebau. So impressed was Riebau with his charge's skill and seriousness of purpose that he dashed off a letter to a London magazine putting Faraday's example forward as a model for the city's youth. (Riebau didn't have much luck bringing his apprentices into the fold: of Faraday's coworkers, reads one account, William Oxberry became a comedian and Edward Fitzwilliam a professional singer.)

Riebau's shop proved a fertile environment for the inquisitive, but virtually unschooled, Faraday. Books came in, books went out, a steady stream of treacle and treasure that Faraday sampled haphazardly in his off-hours. This week's "lesson" might be the *Arabian Nights;* next week's, a collection of Hogarth illustrations; and after that, Fanny Burney's edgy take on English society, *Evelina.* But it was books on science that excited Faraday the most.

At the dawn of the nineteenth century, science and its institutions were in flux, spurred as much by new discoveries as by the growing belief that scientific research might enhance a nation's agricultural and industrial development. The fundamental building blocks of matter—atoms—were as yet unknown. Electricity, magnetism, heat, and light were variously "explained," none convincingly. Through careful measurement, the mathematical character of nature's forces could be determined, but their underlying mechanisms, interrelationships, and means of conveyance through space were subjects of dispute. Faraday plunged headlong into this mélange of ideas, trying with his meager knowledge to sort out fact from fancy. All around was God's handiwork, in plain sight, yet inextricably bound up in mystery, a seemingly limitless horizon of possibilities for off-hours study. Riebau described his apprentice as perpetually scouring the countryside, "searching for some Mineral or Vegitable curiosity . . . his mind ever engaged."

Faraday's scientific musings tumbled joyfully, almost uncontrollably, in his head. In a letter to his best friend, Benjamin Abbott, we see an energetic twenty-year-old dancing through one rain-splashed evening with the unfettered exuberance of a Gene Kelly: "I set off from you at a run and did not stop until I found myself in the midst of a puddle and a quandary of thoughts respecting the heat generated in animal bodies by exercise. The puddle however gave a turn to the affair and I proceeded from thence deeply immersed in thoughts respecting the resistance of fluids to bodies precipitated in them . . . My mind was . . . suddenly called . . . by a very cordial affectionate and also effectual salute from a spout. This of course gave a new turn to my ideas and from thence to Black Friars Bridge it was busily bothered amongst Projectiles and Parabolas. At the Bridge the wind came in my face and directed my attention . . . to the inclination of the pavement. Inclined Planes were then all the go and . . . on the other side of the Bridge . . . slipping introduced the subject of friction and the best method of lessening it . . . The Velocity and Momentum of falling bodies next struck not only my mind but my head, my ears, my hands, my back and various other parts of my body . . . [F]rom thence I went home, sky-gazing and earnestly looking out for every Cirrus, Cumulus, Stratus, Cirro-Cumuli, Cirro-Strata and Nimbus that came above the horizon."

From the start, Faraday's investigations were more than a joyous commune with nature; they were a sincere attempt to discern God's invisible qualities through the very design of the world. Through well-constructed observations and experiments, he sought to distill nature's seemingly diverse phenomena to a common, irreducible basis—and in this fundamental unity of the universe, he would witness the divine signature. The intense spirituality that imbued Faraday's science derived from his upbringing in the Sandemanian faith, a tightly knit Protestant sect founded in the mid-1700s by Scottish minister John Glas and his son-in-law Robert Sandeman. Inspired by a literalist reading of the New Testament, Sandemanians eschewed pride and wealth in favor of piety, humility, and community with fellow Sandemanians. Much of Faraday's overt serenity owed itself to the affirmative aspects of his religion. "He drinks from a fount on Sunday which refreshes his soul for a week," noted friend and biographer John Tyndall. Faraday, the Sandemanian, took human fallibility as a given, so he never staked his ego on the correctness or acceptance of his own ideas. He was a scientific pilgrim, inching his way toward the heart of a complex universe. Whether his chosen path proved mistaken was of little consequence; there was always another path. The joy was in the journey.

Although Faraday might have fancied himself a protoscientist, he was too grounded not to see who stared back at him from the mirror: a rough-edged, ill-educated son of a blacksmith. He needed a mentor—a Henry Higgins, really—to smooth the edges and guide his further education. The "mentor" arrived at Riebau's door in 1809—not surprisingly, in the form of a book.

In the opening paragraphs of *Improvement of the Mind,* the Reverend Isaac Watts spoke directly to young Michael Faraday: "Even the lower orders of men have particular callings in life, wherein they ought to acquire a just degree of skill." Preacher, writer, and composer of hymns (most notably, "Joy to the World"), Isaac Watts labored twenty years on his commonsense self-help guide before it was published in 1741. Faraday read the 1809 edition of *Improvement of the Mind* cover to cover and, as Riebau noted, often carried it in his pocket. In this single volume was a system by which he might organize his swept-up jumble of facts and observations about nature, a system that might help him penetrate the more rarefied strata of English society. It was Watts, Faraday once told a friend, who first made him think.

In unwavering conviction, Watts dispenses advice and encouragement on every facet of self-improvement, from attending lectures to conversation to meditative study. "Do not content yourselves with mere words and names," Watts counsels, "lest your labored improvements only amass a heap of unintelligible phrases, and you feed upon husks instead of kernels." He promotes the importance of direct observation and cautions against the use of imprecise language, twin credos to which aspiring scientist Faraday adhered with almost biblical devotion. And although Watts may have been his readers' most ardent champion, he reins in their wilder aspirations: "Let not young students apply themselves to search out deep, dark, and abstruse matters, far above their reach, or spend their labor on any peculiar subjects, for which they have not the advantages of necessary antecedent learning, or books, or observations."

Faraday hurled himself into Watts's self-improvement plan. Characterizing his own language as "that of the most illiterate," he took elocution lessons two hours a week for seven years. He ordered his friends to mercilessly correct his speech, spelling, and grammar. He began a commonplace book titled "The Philosophical Miscellany," whose pages he filled with facts about light and color, electric fish, meteorites, lightning, water spouts, the formation of snow, loosening glass stopples, oxygen gas, galvanism. He took drawing lessons from French artist Jean-Jacques Masquerier, who had fled Paris in 1792 and lodged in a room above Riebau's shop.

In early 1810, Faraday was binding a volume from the 1797 edition of the *Encyclopaedia Britannica.* Poring through it, he discovered the 127-page entry on electricity. The author was the notorious Scottish surgeon and inventor James Tytler, whose many exploits include the first manned hot-air balloon ascent in Britain—a feat he then repeated to a paying crowd. (The Montgolfier brothers had ascended in France a year earlier, in 1783.) In their acknowledgments, the *Britannica*'s editors refer drolly to Tytler as "a man who, though his conduct has been marked by almost perpetual imprudence, possesses no common share of science and genius."

According to Tytler, everyone agreed that electricity exists in two varieties, which Benjamin Franklin had termed "positive" and "negative." Further, no one disputed that the transfer of electric charge, whatever it was in fact, could be envisioned as the flow of an imponderable fluid that coursed freely through ordinary matter; or that these electrical fluids somehow influenced one another at a distance. Here the agreement ended. French scientists promoted a two-fluid theory, one positive and the other negative, flowing simultaneously in opposite directions. English scientists generally supported Franklin's one-fluid theory, in which electrification results from a surplus or deficit of a single type of charge. (Franklin guessed that the mobile charge was positive; in fact, it turned out to be negative.) Tytler, in the *Encyclopaedia,* put forward a fanciful variation of Franklin's single-fluid theory that not only explained all electrical phenomena—at least in his own inflated opinion—but linked electricity to light and heat. To Faraday, the notion that learned minds might disagree about the nature of such fundamental phenomena as electricity, light, and heat must have been—well, electrifying.

From page 119 of *Improvement of the Mind,* Faraday's "mentor," Isaac Watts, whispered into his ear about "the importance of the observed fact." Do not merely read about electrical phenomena; witness them firsthand. For seven pence, Faraday bought a pair of glass jars from a secondhand shop on Little Chesterfield Street. The larger jar became the core of an electrostatic generator—a device that uses friction to build static electricity—and the smaller one a Leyden jar, to store the electrical charge produced by the generator. (In my own introductory physics classes, students make the generator out of a nested Styrofoam dinner plate and aluminum pie pan, and the Leyden jar out a foil-wrapped, water-filled, film canister with a protruding nail.) With his crude apparatus, Faraday electrified common objects, zapped his outstretched finger with sparks, and tasted electricity's sting on his tongue. No doubt he performed these experiments for Riebau and subjected him to monologues about the one-fluid-versus-two-fluid debate. Riebau could only listen admiringly to the tale of a science disunited. As to how his energetic apprentice might resolve such an issue, he had no idea. But Isaac Watts did, right on page 33 of *Improvement of the Mind:* "There is something more sprightly, more delightful and entertaining in the living discourse of a wise, a learned, and well-qualified teacher, than there is in the silent and sedentary practice of reading." On Monday evening, February 19, 1810, Faraday took a borrowed shilling and marched through central London to his first lecture on science.

Faraday's London bubbled with scientific and pseudoscientific spectacle. Astronomy, meteorology, chemistry, optics, electricity— the stuff of wonder that drew both king and commoner to exhibition halls throughout the city. On any given evening, audiences could thrill to lightning bolts at the Theatre of Science on Pall Mall; watch the guillotined "head" of the great French chemist Antoine Laurent Lavoisier materialize from the mists of the Phantasmagoria at the Lyceum; or witness hair-raising fulminations of the "devil's element," phosphorus, at the Royal Institution on Albemarle Street. On lecture nights, the convergence of carriages was so dense in front of the Royal Institution that the street had to be designated one-way. As the great demonstrators entertained awestruck throngs, their more reserved counterparts lectured at hospitals, private homes, rented halls, and philosophical societies. Citizens excluded from formal schooling could attend these ad hoc "classrooms" and hear about developments in science. Increasingly, knowledge and career choices, technology and societal well-being were linked in the minds of artisans like Faraday.

Taking paper and pencil, Faraday set out for the evening lecture at John Tatum's house on Dorset Street near Salisbury Square. The storefront advertisement had suggested at least a workmanlike recitation of facts and principles on a variety of subjects: fluids, optics, geology, mechanics, chemistry, astronomy, meteorology. But there were also to be demonstrations, likely with apparatus he could ill afford. Faraday was already versed in this evening's topic–electricity–from Tytler's entry in the *Encyclopaedia*. Arriving at Tatum's, he settled into a front-row seat, placed his hat on his knees, and laid a sheaf of paper on top, looking very much the incarnation of Isaac Watts's ideal pupil.

That the lecture would be delivered by Tatum himself–a silversmith, not a "bona fide" scientist–was no surprise. The professional ranks were largely closed to those outside the Cambridge-Oxford axis–unless, of course, one were wealthy or titled. Tatum, like many others at the time, had forged his own parallel universe of science: conducting experiments, giving lectures, even founding a modest scientific organization, the City Philosophical Society, which Faraday subsequently joined. The CPS met every Wednesday and comprised a cross-section of trades: clerk, warehouse worker, landscape painter, solicitor, pharmacist, minister, medical student–all of whom shared an abiding interest in contemporary science. A more like-minded group Faraday could not have found if Isaac Watts had conjured it up himself. For harmony's sake–and in deference to that era's politically repressive environment–only two discussion subjects were off-limits: politics and religion. (The CPS was briefly banned in 1817 under the Seditious Meetings Act.)

Faraday recorded the essence of Tatum's electricity lecture and sketched the various pieces of demonstration equipment. When he arrived home, he composed a second, more detailed account based on his lecture notes, and over the succeeding days, transcribed these into an extended narrative, appending his own opinions on the subject. With each succeeding Monday lecture, he added another chapter to what would become his own sourcebook on the sciences. When it was all done, he bound the volumes and dedicated them to his master, George Riebau. "To you . . . ," he wrote, "is to be attributed the rise and existence of that small portion of knowledge relating to the sciences which I possess . . ."

In the matter of electricity, Tatum was an avowed Franklinian—a "one-fluid" man. Faraday's take on the evidence was different. He set aside the various theories of electricity advanced by Benjamin Franklin, James Tytler, and the French in favor of an obscure two-fluid model developed in the 1770s by Henry Eeles. No matter that England's venerable Royal Society, bulwark of the science establishment (and not affiliated with the Royal Institution), had repeatedly spurned the work of Eeles. Faraday was convinced that it better explained electrical phenomena. Evidently, Faraday made his views known to Tatum, for several weeks later he was standing at the lectern delivering his first public lecture. He left nothing to chance: His lecture notes form a precise script, from the opening "Ladies and Gentlemen" to the extended passages lifted from Tytler. As to why his favorite theory of Eeles had not come to the fore, Faraday had prepared a volley against the "Spirit of Party and bigotry which is to be found as much among *Philosophers* [i.e., scientists] as among *Politicians* and enthusiasts." But he never fired the shot; the words are crossed out in his notes. Instead, Faraday performed demonstrations in support of the Eeles model: passing a spark through a suspended stack of paper; administering electric shocks of increasing intensity to his own body; observing electrical discharges in a partly evacuated tube. In truth, the demonstrations were inconclusive; Franklinians and Frenchmen could spin them to their own advantage.

If Faraday basked in the glow of his first public lecture, it was only briefly. Tatum followed up with a presentation that revealed to Faraday the backward state of his knowledge. Faraday's electrical "bible"–Tytler's *Encyclopaedia* article–dated from 1797. Its conclusions were drawn solely from observations of *static* electricity, the transient rush of surplus electric charge from one body to another. Whatever Faraday had gleaned from Tytler placed him, unwittingly, at least a decade behind the times. Now Tatum described Alessandro Volta's revolutionary invention in 1800 of the battery, the first means of *continuous* electrical flow. Then he spoke about the burgeoning field of electrochemistry, in which electric currents are used to probe the structure of matter. He gave accounts of observations and experiments not even contemplated in Tytler's dated treatise. One thing Faraday learned from Tatum's lecture: It was time to find a newer book.

Among the works that gravitated to Riebau's shop–and into Faraday's hands–in early 1810 was one of the most enduring science books of the nineteenth century. First published four years earlier, Jane Marcet's *Conversations on Chemistry* went through eighteen British, four French, and twenty-three American printings. Born to a wealthy Swiss merchant family living in England, Marcet was educated by private tutors in mathematics, astronomy, philosophy, and the arts. When she was thirty, she married a fellow Swiss, who was a physician and chemistry lecturer at Guy's Hospital in London. Marcet took an active interest in her husband's work and became a familiar figure within his professional and social network. To their scientist friends, she was the ideal conversation partner, ever eager to hear the latest theory or experimental finding. Marcet became a regular at scientific lectures, most notably those of England's celebrated chemist, Humphry Davy. It was the charismatic Davy who inspired Jane Marcet to write her layperson's guide to chemistry. And it was her book, in turn, that inspired Michael Faraday to focus his wide-ranging interests on that subject.

Conversations on Chemistry features a series of dialogues between a teacher, Mrs. B., and her two pupils, sober-minded Emily and impetuous Caroline. The book's structure echoes that of Galileo's *Dialogue on the Two Chief World Systems,* which portrays the discourse among a knowledgeable tutor, an inquisitive nobleman, and a dimwitted Aristotelian who repeatedly stumbles over his own illogic. In the Preface, Faraday read about Marcet's presence at lectures, her scientific conversations, her reliance on proof by experimentation—all familiar elements of self-improvement à la Isaac Watts. Marcet's alter ego, Mrs. B., "spoke" to Faraday in revelatory language that promised fresh insight into the mysteries of nature's "laboratory"—the universe. "I assure you," she said, "that the most wonderful and the most interesting phenomena of nature are almost all of them produced by chemical powers." She buttressed that promise, not with conjecture or blind allegiance to tradition, but with proven facts. Further, Marcet explicitly confirmed Faraday's notion of science as a spiritual exercise—in her words, "a lesson of piety and virtue." In Jane Marcet, Faraday had found both teacher and kindred spirit. And like the book's fictional Emily, he became a diligent pupil.

"Do not suppose that I was a very deep thinker or was marked as a precocious person," Faraday wrote many years later. "I was a very lively, imaginative person, and could believe in the Arabian Nights as easily as in the Encyclopaedia. But facts were important to me and saved me. I could trust a fact, but always cross-examined an assertion. So when I questioned Mrs. Marcet's book by such little experiments as I could perform, and found it true to the facts as I could understand them, I felt that I had got hold of an anchor in chemical knowledge and clung fast to it."

Conversations in Chemistry covers a diverse array of "chemical" topics, reflective of the era: individual elements, light and heat, metals, acids, pharmaceuticals, geochemistry, fermentation, and living organisms. Marcet describes experiments that anyone, using common household materials, can perform at home or in a classroom. Educational reformers, especially at women's academies in the United States, preferred her focus on theory and experiment to the various metaphysical, religious, or narrowly utilitarian approaches of competing texts. On the controversial issue of chemical affinity—the tendency of certain substances to combine in chemical reactions—Marcet allies herself with her friend, Davy. In their view (not the majority opinion at the time), chemical affinity is based on matter's inherent electrical properties. The notion that electricity is a phantom fluid unlinked to its host matter is countered, they believed, by a simple observation: When a battery's electricity is applied through a pair of immersed electrodes—conductors connected to the battery's positive and negative terminals—water splits into its constituent hydrogen and oxygen. Further, the sundered elements appear at opposite electrodes, reflecting their innate electrical charge.

Marcet's electrochemical tales turned Faraday into an amateur chemist—and a Humphry Davy devotee. Davy had risen from modest roots and now held an esteemed position at the Royal Institution on Albemarle Street, barely a mile from Riebau's shop. What a thrill it would have been for Faraday to catch a glimpse of the famed chemist, maybe even exchange greetings—or, better yet, hear from the man himself about the particulars of his research. But

Faraday knew that in early nineteenth-century England the distance between a bookbinder's apprentice and a renowned scientist was vastly greater than a few city blocks. His hero, Davy, might as well have been on the far side of the earth.

One winter evening in 1812, George Riebau retrieved the notes of Tatum's lectures that his apprentice, Michael Faraday, would someday bind and dedicate to him. He must have felt a paternal pride carrying the thick stack of pages, carefully handwritten and illustrated by one of his own. Riebau always encouraged Faraday's scientific activities—as remote as these were from the needs of his own business. He often lent him books from his private collection and shooed the other apprentices out of the backroom "laboratory." This evening, Riebau displayed Faraday's work to one of his customers, a Mr. Dance, who belonged to a prominent London family and was a member of the Royal Institution. Impressed, Dance arranged for Faraday to be admitted to the Institution for the series of farewell lectures by Humphry Davy. The nation's most acclaimed communicator of science to the masses, Davy was scheduled to speak only four more times, between February and April 1812. After that, he would devote himself exclusively to research and travel.

Scottish critic Thomas Carlyle described the Royal Institution, where Davy lectured, as "a kind of sublime *Mechanics' Institute* for the upper classes." Davy's lectures had not only elevated science in the minds of Londoners, it made the acquisition of scientific knowledge fashionable. His audience included government ministers, politicians, dignitaries from overseas, and a significant number of young women. For many of the latter, the draw was the poetic, handsome Davy himself. Faraday, of course, was well acquainted with Davy's electrochemical discoveries, courtesy of Jane Marcet.

But now, unexpectedly, he would have the chance to hear of these and other experiments from the very source.

Just before 8 P.M. on February 29, 1812, Faraday joined the throng of some seven hundred well-heeled visitors squeezing through the front door of the Royal Institution. He crossed the vestibule and climbed the elegant stone staircase to the lecture hall. He sat in the gallery, as befitted his station. Bookbinders, invited or not, never ventured onto the main floor, which was reserved for the upper classes and distinguished guests. From his seat, just over the clock, he had a clear view of the large U-shaped table up front, laden with Davy's equipment. Behind the table were a blackboard and furnace. And in the basement lay the world's most powerful battery, whose electricity Davy tapped from wires that rose through the floor. Looking around at the crowded tiers of seats, Faraday must have thanked providence for his good fortune. Entry to Davy's lectures was by subscription, affordable only to those with means. His own budget strained at the shilling fee for John Tatum's lectures. Nevertheless, here he was, in this exalted institution, in the presence of his hero. As Davy began to speak, Faraday feverishly took notes.

Davy lectured from memory, his script and performance as carefully prepared as an actor's. The presentation was equal parts instruction, performance, and inspiration. Every theoretical assertion was backed by an experiment that he demonstrated on the spot. Davy was animated by the joy of discovery. He delighted in presenting controversial issues, making his argument in a crescendo of logic and experiment. Look what I have found, he seemed to say to the audience, marvel at how it fits into the divine pattern, how it reveals "the power, wisdom, and goodness of the Author of nature." Davy's distrust of speculation echoed Faraday's own. Hypotheses are, as Davy had previously written, "mere points for employing the lever of experiment." The advancement of truth inevitably requires their destruction. Theory is mutable; only facts are eternal.

The Royal Institution of Great Britain circa 1840.

In his final series of lectures, Davy brought his electrochemical wizardry to bear on a seemingly simple question: What is an acid? Acids, such as lemon juice or vinegar, had long been recognized for their common properties: They taste sour (the term *acid* comes from the Latin *acere*, or sour); are corrosive to metals; and give a reddish tinge to litmus, a dye extracted from lichens. The existing chemical paradigm—that oxygen renders compounds acidic—was established during the previous century by Antoine Laurent Lavoisier, who had already proven oxygen's key role in combustion (burning). Davy disputed Lavoisier's claim. He had applied powerful electric currents to muriatic acid, and transformed it into a pungent, greenish gas. Most chemists believed the gas to be an oxygen compound, but in years of laboratory analysis Davy was unable to detect any sign of oxygen. He instead determined that the gas was a combination of hydrogen and a new element, which he named chlorine. Lavoisier's acid theory, in Davy's view, was unfounded. It was not the presence of oxygen, but some overall electrical property of a compound, that gives rise to acidic properties. (Specifically, when dissolved in water, acids release positive ions—atoms or groups of atoms that are electrically positive. Other substances, called bases, release negative ions when they are dissolved in water.) The chemical establishment castigated Davy, accusing him of jumping to conclusions and using faulty lab technique. Oxygen, they believed, would eventually be found by other more capable chemists. In Davy's four lectures, Faraday heard his hero detail the case for the overthrow of Lavoisier's theory. Afterward, there was no doubt in Faraday's mind that Davy was right about acids—and that electrochemistry was a promising path of research.

In July 1812, having saved up money for proper supplies, Faraday decided to launch his own electrochemical experiments. First, he needed a battery. From all he had read and heard, constructing a battery was straightforward. Alessandro Volta had been clear twelve years earlier when he notified Joseph Banks, president of England's Royal Society, of his invention of the battery:

"The apparatus . . . which will, no doubt, astonish you, is only the assemblage of a number of good conductors of different kinds arranged in a certain manner. Thirty, forty, sixty, or more pieces of copper, or rather silver, applied to a piece of tin, or zinc, which is much better, and as many strata of water, or any other liquid which may be a better conductor, such as salt water, ley, &c. or pieces of pasteboard, skin, &c, well soaked in these liquids, such strata interposed between every pair or combination of two different metals in an alternate series, and always in the same order of these three kinds of conductors, are all that is necessary for constituting my new instrument."

Volta was equally voluble about the physiological effects of such chemically generated electricity, which he applied liberally to various parts of his body. To the fingers: ". . . a very disagreeable quivering and pricking." To the mouth: ". . . a sensation of light in the eyes, a convulsion in the lips, and even in the tongue, and a painful prick at the tip of it, followed by a sensation of taste." To the ears: ". . . a kind of crackling with shocks, as if some paste or tenacious matter had been boiling . . . The disagreeable sensation, and which I apprehended might be dangerous, of the shock in the brain, prevented me from repeating this experiment."

Volta's letter was published, and scientists all over Europe were soon building their own batteries. In England, William Hyde Wollaston made a simple voltaic "pile" by stacking English shillings, zinc, and pasteboard. The noted chemist Jöns Jakob Berzelius in Sweden couldn't afford silver so he used disks of copper instead. And an up-and-coming chemist named Humphry Davy tried his own hand at battery construction at Bristol's Pneumatic Institution, where he was employed. By the time Michael Faraday marched down to Knight chemists on Foster Lane to buy a sheet of zinc in 1812, anyone with a mind to could assemble a battery. (You can make a battery from a Dagwood "sandwich" of alternating pennies, dimes, and moistened paper; its feeble power won't light a bulb, but will run a small calculator.)

"I, Sir, I my own self, cut out seven discs [of zinc] of the size of half-pennies each!" Faraday wrote his friend Benjamin Abbott on July 12, 1812. "I, Sir, covered them with seven halfpence and I interposed between seven, or rather six pieces of paper soaked in a solution of muriate of soda [i.e., salt water]!!! But laugh no longer, dear A., rather wonder at the effects this trivial power produced." (From a modern perspective, chemical reactions between the metal electrodes and the fluid, or electrolyte, create a surplus of negatively charged electrons at one electrode and a dearth at the other. Thus, one battery terminal becomes the "negative" and the other the "positive." This electrical imbalance impels electrons within an external wire connecting the terminals to flow, as long as the chemical reaction proceeds.)

Faraday next attached copper wires to the ends of his battery-stack, dipped these into a dissolved solution of Epsom salts—magnesium sulfate—then watched in amazement as the electricity did its work. "[B]oth wires became covered in a short time with bubbles of some gas, and a continued stream of very minute bubbles, appearing like small particles, ran through the solution from the negative wire. My proof that the sulphate was decomposed was, that in about two hours the clear solution became turbid: magnesia was suspended in it." Somehow the electrical influence had separated the dissolved magnesium sulfate into its constituent parts, just as Marcet and Davy had asserted.

Fired up by his initial success, Faraday returned to Knight's for more zinc and, this time, a sheet of copper. From each metal, he cut out twenty disks, about 1¾ inches in diameter. These he layered with flannel scraps moistened with salt water. He applied this stronger battery to a number of compounds: magnesium sulfate again, copper sulfate, lead acetate. All decomposed into their constituent parts. Then he tried to decompose water from the shop's lead cistern. The result—a white precipitate—was at odds with what he had read. The anomalous outcome he attributed (probably correctly) to impurities in the water: lead, iron, salt, carbon dioxide.

But, Faraday informed Abbott, there was more. "[O]n separating the discs [of the battery] from each other, I found that some of the zinc discs had got a coating . . . of metallic copper, and that some of the copper discs had a coating of the oxide of zinc. In this case the metals must both have passed through the flannel disc holding the solution of muriate of soda, and they must have passed by each other." What, Faraday wondered, was the precise action that ripped these metals from their native plate, then guided them to their destination? And if, as conventional wisdom held, the mutual force of the zinc and copper plates plucked particles from each other, and if these migrating particles were oppositely charged—again mutually attractive—how did the particles manage to slip past one another without combining? By the time Faraday teased out the answers to these questions, decades hence, he would have completed his own journey, guided by the inexorable force of curiosity.

As remarkable as his experiments might have seemed to him at the time, it must have crossed Faraday's mind what he might accomplish with more than the "trivial power" at his disposal. He knew that the Royal Institution had provided Davy with powerful banks of batteries, including one with two thousand pairs of metal plates that took up an entire bunker in the basement. Shocking various solutions with these, Davy had isolated pure forms of elements, both known and unknown at the time: sodium, potassium, calcium, magnesium, barium, strontium, and chlorine. Connecting his mega-battery to a pair of carbon electrodes, Davy had also created the first continuous electric light, which blazed as brightly as the sun, according to one witness.

What might have appeared as magic to most, Faraday recognized as the judicious use of equipment—admittedly in very capable hands. He had carried out basic chemical experiments and now wished to go further. However, he was trapped by his own procedural philosophy: his need to prove—or disprove—assertions through experiment. "I was never able to make a fact my own without seeing it," he wrote to a friend. "[H]ow terrified I should be to set about learning science from books only." His experimental aspirations required a true laboratory, not a mantelpiece toy. As long as he was a bookbinder, he would have neither the time nor the means to pursue science the way he wanted.

As the end of his apprenticeship loomed, a career in the sciences seemed increasingly remote, and "the binding of other men's thoughts in leather backs, seemed the only means of livelihood open to him." In desperation, he wrote to Joseph Banks, president of the

Royal Society, begging for a scientific position, "however menial." He checked back with the porter several days later. No answer. He returned three more times. Still no response. Finally, the porter delivered Banks's reply—"the letter required no answer." In the eyes of the scientific establishment, Michael Faraday did not exist.

2

PERCEPTIONS PERFECTLY NOVEL

When the mind is ready, a teacher appears.

—CHINESE PROVERB

England's county Cornwall juts boldly into the Atlantic at the island's southwest extremity. This haunted landscape of moors, cliffs, and beaches lies about as far as one can get from the rattle and hum of London without wading into the sea. Its secluded coves once sheltered a ragtag navy of privateers who struck at passing vessels. It's not surprising that the coastal town of Penzance, nestled picturesquely against Mounts Bay, is known more for its fictional pirates than for the famous flesh-and-blood scientist who was born there.

Eldest of five children of a wood carver and his milliner wife, Humphry Davy showed intellectual promise from an early age. In grammar school, he regaled his fellow students with ghost stories and tales from the *Arabian Nights,* and by his teenage years was already a practiced chemist. Among his homemade creations was "thunder powder," which he routinely exploded for his friends. He tramped the Cornish countryside, fishing, hunting, collecting rocks, and otherwise finding inspiration for his poetry.

Davy's high-spirited romp through childhood ended abruptly at age sixteen when his father died. Apprenticed by his godfather in 1795 to a local apothecary-surgeon, Davy appeared headed for a medical career. His diary lists an ambitious plan of self-study indicative of his intellectual energies: theology, geography, logic, languages (seven in all, including Hebrew), physics, mechanics, rhetoric and oratory, history and chronology, mathematics, plus a host of medical-related fields. His increased access to chemical supplies stoked a simmering interest in experimentation—specifically, in chemistry. Through chemical manipulation, he developed his own pigments for painting and analyzed the makeup of local seaweed. After a friend took him to see a chemistry laboratory in Hayle, Davy realized that a life as a country practitioner was not for him.

Davy's entry into the world of chemistry couldn't have been more timely. The investigation of matter's innate properties had thrown off its long-held ties to medicine and pharmacy and established itself as a field in its own right. Nations turned to chemistry for solutions to practical problems in manufacturing, agriculture, and warfare. Davy's primary resource was Antoine Laurent Lavoisier's *Elementary Treatise on Chemistry*. More than anyone else, Lavoisier transformed chemistry from a magical art into a modern science. Lavoisier proved that fire is a form of energy derived from chemical reactions of substances with oxygen. He demonstrated that water, previously believed to be an indivisible element, is a compound of hydrogen and oxygen. He further posited that the total quantity of matter before and after a chemical reaction is the same, although some fraction might change form. "Upon this principle," Lavoisier wrote, "the whole art of performing chemical experiments depends." In Lavoisier's zero-sum game—mass in equals mass out—chemists were compelled to quantify every product of a chemical reaction and to account for any "missing" matter. Finally,

Lavoisier constructed the first comprehensive and essentially modern list of chemical elements, erring only with his inclusion of several compound substances plus light and heat.

Lavoisier's theory of heat was controversial from the moment he proposed it. In the great chemist's own book, Davy read that heat is a material essence—a fluid called caloric—that substances absorb or give out; thus, water absorbs caloric from a flame and becomes hot. However, in William Nicholson's *Dictionary of Practical and Theoretical Chemistry,* Davy found an alternative view: Heat is a form of energy—the collective vibrations and motions of matter's constituent particles. To decide which theory was correct, Davy conducted a series of crude experiments in 1798 in which he melted pieces of ice and wax by friction. The results, in Davy's view, negated Lavoisier's caloric theory. Davy published his findings, along with a raft of ill-conceived speculations about light and matter. The ensuing criticism from the scientific community grounded Davy's youthful flights of fancy. "I was wrong in publishing a new theory of chemistry in such haste," he admitted. "My mind was ardent and enthusiastic. I believed I had discovered the truth." Still his ingenuity shines and foreshadows the brilliant experimenter he would become. (Simultaneously, in Bavaria, American military advisor Benjamin Thompson—Count Rumford—administered the death blow to the caloric theory. Rumford noted that in the boring of cannons, the amount of heat generated by friction is virtually limitless, instead of declining with time as caloric is used up. Further, the total mass of a cannon and its filings remains the same despite the supposed infusion of prodigious amounts of caloric.)

In 1799, nineteen-year-old Davy left his apprenticeship for a position at the newly established Pneumatic Institution in Bristol. Founded by the erratic Thomas Beddoes, a former Oxford chemistry professor dismissed for his liberal political views, the Pneumatic Institution investigated possible therapeutic benefits of various gases for "persons with Consumption, Asthma, Palsy,

Dropsy, obstinate Venereal Complaints, Scrofula or King's Evil, and other diseases which ordinary means have failed to remove." Beddoes was amply funded by wealthy donors, such as James Watt and Josiah Wedgewood, whose relatives were afflicted with tuberculosis. He lavished Davy with laboratory equipment and supplies, not just to produce and analyze gases, but to support other lines of chemical research that might prove lucrative (such as electrochemistry). Beddoes also subscribed to all the professional journals, so his able assistant might keep abreast of important developments in the field. In the meantime, Beddoes tended to patient care and self-promotion.

At the turn of the eighteenth century, the only means to assess the effects of a gas on the human body was to breathe it. This, Davy carried out with utter fearlessness. He nearly asphyxiated himself on several occasions, inhaling gases now known to be toxic, such as carbon monoxide and carbon dioxide (although he ventured that diluted carbon dioxide promotes sleep). Nitric oxide, he noted, "produced a spasm of the epiglottis so painful as to oblige me to desist instantly. When I opened my lips to inspire common air, nitric acid was instantly formed in my mouth, which burnt the tongue and palate, injured the teeth, and produced an inflammation of the mucous membrane which lasted some time."

Davy's most productive trials were with nitrous oxide—laughing gas—believed by some to be poisonous. Breathing it in, he experienced a drunken giddiness. "A thrilling extending from the chest to the extremities was almost immediately produced. I felt a sense of tangible extension highly pleasurable in every limb; my visible impressions were dazzling and apparently magnified, I heard every sound in the room and was perfectly aware of my situation. By degrees as the pleasurable sensations increased, I lost all connection with external things; trains of vivid visible images rapidly passed through my mind and were connected with words in such a manner, as to produce perceptions perfectly novel." Awakened from his trance, Davy exclaimed, "Nothing exists but thoughts!" He

Humphry Davy, engraving from the original Royal Society portrait.

breathed nitrous oxide three or four times a day for several months to assure himself of its safety. He reported his experiences to Beddoes, who tried the gas on a young female patient. To everybody's astonishment, the "fair fugitive, or rather temporary maniac," dashed out of the building, jumped over a dog, and raced across the square before being captured. Davy offered the gas to his friends, among whom were such literary notables as Robert Southey and Samuel Taylor Coleridge. Afterward, Southey wrote to his brother, "Davy has actually invented a new pleasure, for which language has no words . . . I am sure the air in heaven must be this wonder-working gas of delight!" Not everyone was so enamored of the Pneumatic Institution's activities: One magazine castigated the "Pneumatic Revellers" of Bristol for indulging in the mind-altering gas.

Word of Davy's inhalation experiments spread rapidly, at first through correspondence and gossip, then through pamphlets and journal articles. Soon Davy's name was known throughout Europe as the expert on nitrous oxide. One enterprising marketer offered green silk inhalation bags, modeled after the one used by Davy. In 1800, Davy published a 580-page tome, titled *Researches, Chemical and Philosophical, Chiefly concerning Nitrous Oxide . . . and Its Respiration.* His exacting laboratory technique was evident in the detailed narrative, although the public spotlight shone almost exclusively on the candid descriptions of nitrous oxide's exhilarating effects on the mind. Davy points out that nitrous oxide seems to ameliorate pain—the eruption of his own wisdom tooth as an example—and may have potential as a surgical anesthetic. (Laughing gas entered use in dentistry as a numbing agent forty-five years later.) At the end of his *Researches,* Davy renounces the therapeutic effects of gases, claiming (correctly) that there is no compelling evidence for it. Thus he divorced himself from the suspect field of pneumatic *medicine* and allied himself with its essentially rational sibling, pneumatic *chemistry.* In the eyes of his fellow professionals, Davy's youthful, pseudoscientific indiscretions were forgiven. The book established his reputation as a first-rate analytic chemist. The eminent Joseph Priestley, discoverer of oxygen, wrote to Davy in 1801: "Sir, I have read with admiration your excellent publications, and have received much instruction from them . . . I rejoice that you are so young a man, and perceiving the ardour with which you begin your career, I have no doubt of your success." Beddoes did not fare as well. He pursued his pneumatic quest to a dwindling constituency. By 1803, the Pneumatic Institution was no more.

The multifaceted Davy maneuvered confidently within both literary and scientific circles. His friends and colleagues promoted him ceaselessly. Southey described him as "a miraculous young man . . . the young chemist, the young everything, the man least ostentatious of first talent that I have ever known." Thomas Poole, a tanner and literary light, commented that Davy's "delight was in his intellectual being. He felt that he had the power of investigating the laws of nature beyond that entrusted to the generality of men." And Coleridge, when asked by his publisher to compare Davy to the young men of London, replied, "Why, Davy could eat them all. There is an energy, an elasticity in his mind which enables him to seize on and analyse all questions, pushing them to their utmost consequences."

In March 1801, Davy left Beddoes to become assistant lecturer in chemistry at the Royal Institution of Great Britain in London. The Royal Institution had been founded in 1799 to promote "the application of science to the common purposes of life." The founders, fifty-eight men of property, believed that a populace educated in science and technology would help foster the country's industrialization, self-sufficiency, and imperial aspirations. And with the French Revolution fresh in mind, institutions such as theirs might stave off any social instabilities fostered by the chasm between rich and poor. A large house was purchased on Albemarle Street, whose previous owner had been killed by a highwayman, and fitted with laboratories, meeting and display rooms, living quarters, a library, and an auditorium. Despite the founders' lofty societal goals, the educational thrust was soon scaled back. The workshops for a planned artisans' school were never built and the model Public Kitchen in the basement was dismantled shortly after it opened. The Royal Institution became a research center devoted largely to the applications of chemistry to agriculture (many of the founders were wealthy landowners). The public education role was confined to small specialty classes for medical students and other preprofessionals and to public lectures. That these activities—especially the lectures—could provide an income stream to support the research programs was of vital importance. The auditorium accommodated up to seven hundred paying customers—if such a crowd could be enticed to come. Although successful at filling the seats, the resident chemistry lecturer, Thomas Garnett, had been at odds with the Institution's managers, and was on his way out. In Davy, whose name was already known to the public, the managers believed they had found the ideal draw.

Davy was ecstatic about his appointment to the Royal Institution. London was, in his view, "grand theatre"—and he was eager to be its rising star. At first, he continued to socialize with his literary and left-leaning friends. (Davy was pragmatically neutral when it came to politics.) Evenings might be spent discussing poetry or societal ills at Old Slaughter's Coffee House on St. Martin's Street. But his newfound fame as a lecturer gave him entrée to London's high society. And once there, he embraced it. Davy's friend, Samuel Purkis, described the public frenzy: "The sensation created by his first course of Lectures at the Institution, and the enthusiastic admiration which they obtained, is . . . scarcely to be imagined. Men of the first rank and talent,—the literary and the scientific, the practical and the theoretical, blue-stockings, and women of fashion, the old and the young . . .—eagerly crowded the lecture-room. His youth, his simplicity, his natural eloquence, his chemical knowledge, his happy illustrations and well-conducted experiments, excited universal attention and unbounded applause. Compliments, invitations and presents, were showered upon him in abundance from all quarters; his society was coveted by all, and all appeared proud of his acquaintance."

Women swooned over the dashing young scientist. One London socialite concluded that "it was impossible to have seen him and have supposed him an ordinary person," characterizing his eyes as "radiant of genius, and the most bright and beaming expression of intellect that can possibly be conceived." Another put it more succinctly: "Those eyes were made for something besides poring over crucibles." When Davy fell ill in 1807, so frequent were inquiries that hourly updates were posted outside the building.

Davy was a dynamo onstage, yet sufficiently eloquent to keep his pyrotechnical demonstrations from degenerating into a dog-and-pony show. He laced his lectures with romantic and metaphysical allusions. "Not contented with what is found upon the surface of the earth," he once rhapsodized about chemistry, "[man] has penetrated into her bosom, and has even searched the bottom of the ocean for the purpose of allaying the restlessness of his desires, or of extending and increasing his power." No wonder Coleridge attended Davy's lectures when he needed, in his words, to renew his stock of metaphors. And no wonder half of Davy's audience was female. To his rhetorical prowess were added thrilling visions of the sublime power of nature: chemical color displays, gunpowder explosions, giant sparks, electric-pen drawings, dazzling carbon-arc light. Could an educational hour be more stimulating?

While Davy's personal charisma might have attracted the ladies, his scientific outlook pulled in the wealthy and powerful. In making his pitch for the Royal Institution's program—and his own—Davy carefully navigated the shoals of political expediency. He reassured the aristocracy—"guardians of civilization and of refinement," as he called them—by endorsing class hierarchy and the uneven distribution of wealth. Yet he also urged them to become "friends and protectors of the labouring part of the community." And what better way than through patronage of science. The results of basic research, he told audiences in soaring language, promised economic benefits to agriculture and industry; indeed, science was essential for progress as well as for social stability. For if the masses, through increased knowledge and sense of self-worth, viewed themselves as playing a useful role in society, envy and dissension would whither. So powerful was this message—and so lucrative were Davy's lectures—that the Royal Institution abandoned its avowed mission to be a training school for the "useful arts" and became instead a chemical research center.

Davy was surely not beloved by all. The right-wing press characterized him as a dandy and his public lectures as frivolous and unprofessional. He was criticized for his support of women's education in science, a hot-button issue at the time. (He did admit that women had no role to play in scientific investigation.) As both a geographic and a class "outsider," Davy's personal life underwent constant scrutiny. Gossip swirled around his speech, his clothes, his politics, his masculinity—and later around his marriage and his knighthood. Despite the complexities of his private life, Davy churned out research breakthroughs at a furious pace, establishing for himself a worldwide reputation.

Davy's initial research mandate was grounded in the practical: to explore the chemical aspects of tanning and agriculture. This he did with characteristic fervor, resulting in his promotion to professor of chemistry in 1802 and election to the Royal Society in 1803. He received the Society's highest award, the Copley Medal, in 1805. To the untrained eye, Davy's rapid-fire laboratory technique might have seemed haphazard, even reckless. His spotted laboratory notebooks bear witness to the chaotic scene. "With all his immense ability," chided one biographer, "he was a man almost destitute of the faculties of order and method." Davy often conducted several experiments simultaneously, flitting from one to another like a frenetic chef. The dizzying pace didn't faze him; he routinely hummed while he worked. But this experimental tumult masked an uncompromising degree of rigor and thoroughness. Davy's chemical prowess became legendary. He verified his various hypotheses with such well-designed experiments that they were virtually unassailable on scientific grounds.

While attending to the Royal Institution's practical vision, Davy nevertheless conducted a parallel line of basic research in electrochemistry, recognizing the battery's vast potential for chemical analysis. Already William Nicholson and Anthony Carlisle had split water into its constituents, hydrogen and oxygen, by means of electricity. But why did the hydrogen bubble up at the negative electrode, oxygen at the positive, and neither in the intervening liquid? Why did traces of other elements appear when water, it was almost universally believed, consists solely of hydrogen and oxygen? The scientific journals bulged with absurd explanations: Electricity is an acid; water is an element, not a compound; hydrogen and oxygen migrate through the external wires, not the water.

Davy waded into the fray with a well-constructed series of experiments completed in 1806. Through meticulous laboratory procedures, he sought to identify all sources of possible contamination. He found that the anomalous presence of acids stemmed from electrical action on various salts leached from the electrodes and the glass container. So he replaced glass containers with ones made of agate or gold, which were boiled for several hours prior to use. Electrodes of nonreactive platinum were adopted. And the water itself was repeatedly distilled in a vessel of silver, until no contaminants remained. When he traced the origin of a still-lingering acid to atmospheric nitrogen dissolved in the water, Davy conducted further electrolysis experiments under evacuated glass domes. In the end, Davy was able to confirm that "water, chemically pure, is decomposed by electricity into gaseous matter alone—into oxygen and hydrogen."

Davy also showed more generally that, while the chemical activity *appeared* to be confined to the electrodes, the solution itself is not passive. Dissolved substances migrate in one direction or another under the influence of the applied electricity. In one case, Davy connected three solution-filled vessels in series by means of porous asbestos wicks. Into one of the outer vessels he dissolved the compound sodium sulfate, and into the other, barium nitrate. When electricity was applied to the outer vessels, a precipitate of barium sulfate formed in the middle vessel. Evidently, barium and sulfate ions had each migrated through the asbestos wicks from their respective host vessels into the middle vessel, where they linked together.

In another experiment, Davy filled two glass tubes with distilled water and ran an asbestos wick from each to a middle tube containing a solution of potassium chloride. Electrodes were inserted into the outer, water-filled tubes. Soon, the water in one of the electrified tubes turned acidic, and the water in the other turned alkaline. Potassium ions had migrated toward the negative electrode, revealing their inherent positive charge; chlorine ions had gone the other way, and thus must be negative. The divergent movements arose even though the electrodes were never in direct contact with the "source" solution in the middle. Electricity's effect occurs, not just at the electrodes, but penetrates throughout a solution.

Davy conducted scores of other electrochemical experiments, highlighting matter's innate electrical properties and how these influence chemical reactions. Before, chemistry was a "kitchen science": There was no way to know whether two substances would react with one another unless you mixed them and noted the outcome. Davy believed that the electrical character of substances, once determined, might be used to predict the results of chemical reactions. Ions with, say, an overall positive charge would have an affinity for those that are negative, while those of the same charge would be nonreactive. To Davy, chemistry was a dynamic of material combination and decomposition, whose arbiter is some innate electrical force within matter itself. And once the *quantitative* electrical properties of matter were referred to the observed proportions in which various substances coalesce or break apart, chemistry would be placed on a more mathematical footing.

Davy's 1806 lecture before the Royal Society, titled *On some Chemical Agencies of Electricity,* summarized his findings and cemented his scientific standing. Despite the ongoing conflict between England and France, Davy received the annual Napoleon Prize from the Institut de France for his electrochemical research. His new electricity-based model of chemistry bore further fruit almost immediately. In 1807, Davy announced the isolation of an entire roster of elements—sodium, potassium, calcium, barium, magnesium, strontium, and chlorine—each one electrically separated from its host compound. The discovery of chlorine was especially significant to the progress of chemistry, for it disproved Lavoisier's generally accepted theory that acids are oxygen-containing compounds. Davy derived his chlorine from the breakdown of muriatic acid—now known as hydrochloric acid—whose sole ingredients, Davy demonstrated, are chlorine and hydrogen. That previous researchers had detected the presence of oxygen in muriatic acid, Davy traced to faulty lab technique: The oxygen likely came from trace amounts of water vapor that had leaked into their reaction vessels. In Davy's opinion, the essential qualities of acids arose from the electrical properties of the acid molecules themselves, not from the presence of oxygen.

Davy's assault on Lavoisier's acid theory triggered a harsh response from Continental chemists. Even the vaunted Berzelius in Stockholm refused to accept Davy's conclusions for another decade. Meanwhile, Davy continued to experiment, although his experiments were now sandwiched into a tightly packed schedule of social engagements, lectures, travel, angling, and hunting. He frequently worked until three or four in the morning, only to rise early to prepare a lecture. To save time, Davy was known to put on fresh clothes over those he was already wearing. (One witness places five shirts on him, and friends used to wonder at his rapid ups and downs in "weight.")

By 1812, knighted and backed by a fortune from his marriage to a wealthy widow, *Sir* Humphry Davy was eager to vacate the now-famous lecture hall at the Royal Institution. He delivered a series of four farewell talks, defending his controversial stance against Lavoisier.

As always, Davy held the audience in enraptured silence, his literate narrative of ideas punctuated by compelling experiments.

Perhaps, as he swept his gaze over the crowd, his eyes lingered now and then on an anonymous face in the gallery, just over the clock. Indeed, the same face in the same place for all four lectures was as predictable as the clockwork above which it inevitably appeared. A young man, eager and attentive, hat perched on his knees, busily taking notes of the proceedings.

On October 7, 1812, Faraday's apprenticeship with George Riebau ended. The following day, he reluctantly took up his job as a journeyman bookbinder with Henri de la Roche, a French emigrant to London. The disagreeable de la Roche was no Riebau when it came to Faraday's scientific pursuits. The bookbinding work was all-encompassing. No more City Philosophical Society lectures. No more backroom experimentation. Just monotony and dim prospects. Faraday wrote to his friend Huxtable on October 18, 1812: "[I] am now working at my old trade, the which I wish to leave at the first convenient opportunity . . . With respect to the progress of the sciences I know but little, and am now likely to know still less."

Just a few weeks later, Faraday found himself on the streets of London in what he might have thought was an impossible waking dream. He was headed toward the Royal Institution, his emotions surely torn between distress at the accident that temporarily blinded Humphry Davy and amazement that the great chemist had called on him—yes, *him*—for help. Davy needed an amanuensis until his eyes healed from the recent explosion of a "detonating compound" he was testing. Riebau's customer, Mr. Dance, having read Faraday's notes of Tatum's lectures, had apparently recommended Faraday for the position. The few days in Davy's presence fired Faraday's hopes for a permanent job at the Royal Institution. In late December, after Davy had recovered, Faraday sent him a letter of inquiry. Accompanying the letter was the best evidence he could muster of his devotion and promise: the cherished volume of notes he had taken of Davy's farewell lectures. Davy responded on Christmas Eve, in a note that Faraday was to keep his entire life:

Sir,

I am far from displeased with the proof you have given me of your confidence, and which displays great zeal, power of memory, and attention. I am obliged to go out of town, and shall not be settled in town till the end of January: I will then see you at any time you wish.

It would gratify me to be of service to you. I wish it may be in my power.

I am, Sir,
Your obedient humble servant,
H. DAVY.

But Davy's power proved insufficient. When the two did meet in early 1813 in the anteroom of the great lecture hall, all Davy could offer was his sympathy. There was simply no place for another assistant at the Royal Institution, no matter how ardent the desire.

Faraday continued to labor dejectedly in his unwanted profession, despite de la Roche's offer to make him heir to the business—a man of property. One evening, not long after his conversation with Davy, a grand coach pulled up to the door of 18 Weymouth Street, where Faraday lived. The footman delivered a note from Davy asking Faraday to come see him at the Royal Institution in the morning. When Faraday arrived, he learned that the Institution's laboratory assistant had been dismissed for brawling with the instrument maker. A replacement was needed immediately. The job was more "chief bottle washer" than chemist—assisting lecturers, maintaining equipment, cleaning display items, sweeping, lighting fireplaces—but to Faraday, it was his long-sought entrée into the scientific world. On March 1, 1813, Michael Faraday joined the Royal Institution, with a salary of 25 shillings per week, lodging in the attic, plus—at Faraday's request—a laboratory apron and permission to use the apparatus for his own experiments.

Faraday was a quick study. Within a week, he was performing routine chemical tasks for Davy. And by April, he was brewing nitrogen trichloride, the fearsome detonating compound that had exploded in Davy's face. (Davy was lucky; the substance's discoverer, French chemist Pierre Dulong, lost an eye and a finger.) Perhaps taking his cue from Davy, Faraday thrilled at the compound's pyrotechnical power. He wrote Benjamin Abbott about igniting samples in a water-filled basin, which invariably shattered with great violence. A subsequent explosion of a tubeful ripped open his hand and numbed his fingers for several days. Davy escaped with a lacerated chin and bruised forehead. Day after day, the two of them charged onto this chemical battlefield, dodging spray and shrapnel to tease out the properties of the temperamental compound. Indeed, Faraday must have felt like a soldier in service to science—he and Davy, together, advancing the front line of knowledge. Faraday told his aunt and uncle, without irony, "I am constantly engaged in observing the works of Nature and tracing the manner in which she directs the arrangement and order of the world."

When not in the lab, Faraday helped set up and conduct lecture experiments for chemistry professor William T. Brande and others. He cast his critical eye on the proceedings, penning to Abbott a virtual manual on the art of the lecture. Excusing his inexperience—"If we never judge at all we never judge right."—Faraday addresses a multitude of practical issues, from the layout, lighting, and ventilation of the hall to the lecturer's bearing and speech. A "good delivery" is essential, he tells Abbott, for "the generality of mankind cannot accompany us one short hour unless the path is strewed with flowers." No doubt Faraday imagined himself standing at the

Royal Institution's U-shaped table addressing an audience. After devoting three lengthy and opinion-filled letters to Abbott on the subject, he raps himself gently for his presumptuousness—before launching into a fourth: "As when on some secluded branch in forest far & wide sits perched an owl who full of self-conceit and self-created wisdom, explains, comments, condemns and orders things not understood: yet, full of his importance still holds forth to stocks and stones around:—so sits, and scribbles Mike."

After work, Faraday read, played his flute, sang (bass) with friends, or just listened to the strains of music that drifted in from the nearby Jacques Hotel. "[T]he music is so excellent, that I cannot for the life of me help running at every new piece they play to the window to hear them." He visited his family frequently on Weymouth Street and, when time permitted, joined his science-minded colleagues at Wednesday meetings of the City Philosophical Society. Faraday's social life, too, revolved around the Society, especially his friendship with Benjamin Abbott. The two met regularly, yet maintained a correspondence in order that they might, in Faraday's words, burnish themselves up a bit. To Abbott, Faraday outlined the touchstones of his own moral life—". . . I keep regular hours, enter not intentionally into pleasures productive of evil, reverence those who require reverence from me, and act up to what the world calls good."—although he did not let his stodgy doctrine stand in the way of fun. Abbott and his brother, Robert, sometimes stopped by the Royal Institution when Davy was away (which was often) to see what Faraday was working on. Occasionally, the three of them engaged in what Abbott termed "little frolics"—inhaling laughing gas or, on one patriotic holiday, setting off firecrackers in crowds of celebrants.

Twenty-fours hours proved insufficient for the tide of Faraday's daily ponderings and activities, a fact he lamented to Abbott: "What is the longest, and the shortest thing in the world: the swiftest, and the most slow: the most divisible and the most extended: the least valued and the most regretted: without which nothing can be done: which devours all that is small: and gives life and spirits to every thing that is great? It is that, Good Sir, the want of which has till now delayed my answer to your wellcome letter. It is what the Creator has thought of such value as never to bestow on us mortals two of the minutest portions of it at once. It is that which with me is at the instant very pleasingly employed. It is Time." Nevertheless, Faraday was all too willing to be interrupted by letters from his friend, even if time constraints forced a telegraphic reply: "–no–no–no–no–none–right–no Philosophy is not dead yet–no–O no–he knows it–thank you–'tis impossible–Bravo. In the above lines, dear Abbott, you have full and explicit answers to the first page of yours dated Septr 28. I was paper hanging at the time I received it but what a change of thoughts it occasioned; what a concussion, confusion, conglomeration, what a revolution of ideas it produced–Oh 'twas too much–away went cloths, shears, paper, paste and brush, all–all was too little, all was too light to keep thoughts from soaring high, connected close with thine."

Faraday was not averse to placing his own psyche under the microscope, and, if apt, applying his observations to the whole of humanity. One solitary evening in his attic room, he jotted a stream-of-consciousness ramble to Abbott, alternately extolling their friendship and sniping at him for writing too infrequently. Instead of throwing it out, as he had planned, he sent the letter to Abbott with this plea for understanding: "What a singular compound is man–what strange contradictory ingredients enter into his composition–and how completely each one predominates for a time according as it is favoured by the tone of the mind and senses and other existing circumstances–at one time grave circumspect & cautious–at another silly headstrong and careless–now conscious of his dignity he considers himself as Lord of the creation–yet in a few hours will conduct himself in a way that places him beneath the level of the beasts—at times free frivolous and open his tongue is an unobstructed conveyor of his thoughts—thoughts which on after consideration makes him ashamed of his former behaviour—indeed the numerous paradoxes anomalies and contradictions in man exceed in number all that can be found in nature elsewhere and separate and distinguish him if nothing else did from every other created object . . ." Although his reply is lost, Abbott accepted the "singular compound" that was his friend Faraday. The two kept up their correspondence into the 1820s and their friendship for life.

By summertime, Faraday's new life had settled into routine. But he was about to learn that life in Davy's orbit rarely stayed routine for long.

His fame and fortune secure, Davy planned to embark with his wife on a multiyear tour of Europe and possibly Asia. Part-sightseeing, part-professional, the trip would include stops at major scientific facilities in France, Italy, and Switzerland, and discussions with renowned scientists there. Davy, of course, would be packing his portable laboratory to conduct experiments on the fly. It was a perilous time for any English citizen, even one of Davy's stature, to travel the Continent. England and France had been at war almost constantly since 1793. And with both Wellington and Britain's allies poised to invade France, the roads were desperate and confused ground, with the promise of danger around every turn. To most, including Faraday, Davy's plan must have sounded like lunacy. But evidently not to Napoléon, who directed that special passports be granted for Davy and his entourage to travel unhindered anywhere within his empire; no one was about to accuse his government of being anything less than progressive toward science. The London press, meanwhile, rebuked Davy for consorting with the enemy.

If Faraday regretted the impending loss of his mentor Davy, it was only for a moment. Davy unexpectedly invited him on the trip in the role of chemical assistant and valet. Faraday weighed the political climate and his deep distaste at becoming a personal servant against the promise of a Continental education. By his own account, he had never strayed more than twelve miles beyond the confines of London. Now his horizons might expand immeasurably. With the worldly Davy as his guide, his tutor, and his ambassador to European society, Faraday would be taking the standard route to refinement of young aristocrats. No classroom, no set of books could duplicate such an experiential education in art, architecture, music, history, languages, social customs—all the while learning analytical chemistry from a master. A dizzying ascent to worldliness for one whose childhood home lay above a stable. " 'Tis indeed a strange venture at this time," Faraday admitted. Yet, he decided, ultimately worth the risk.

On October 13, 1813, Faraday set out with Davy and his wife for the Continent. "This morning," commenced his travel journal, "formed a new epoch in my life."

3

THE UNIVERSITY
OF EXPERIENCE

*Man's mind stretched to a new idea never goes back to its
original dimension.*

—OLIVER WENDELL HOLMES

They were an incongruous foursome, bumping along in the gleaming black carriage toward Plymouth, where the ship to France awaited. Sir Humphry Davy—illustrious poet-scientist, recently gilded by knighthood and marriage; his wife, Lady Jane Davy—socialite and unrepentant snob, eager to frequent the plush salons of the Continent; her maid, the aptly named Mrs. Meek; and, exiled to the roof with the footman and driver, twenty-two-year-old Michael Faraday—chemical assistant and reluctant valet, studying the novel landscape as though it were a continuous miracle. Whatever objections Lady Davy might have expressed to her husband about inviting along this untutored bumpkin were voiced in private. Nevertheless, here he sat atop *her* carriage, with his plain clothes, his plain manner, and that plain face so aptly caricatured in his passport description: round chin, brown beard, large mouth, great nose. Lady Davy was the central sun around which Edinburgh's literary lights revolved, admired to such a degree for her beauty, wit, and personality that a venerable professor had once stooped in the street to adjust the laces of her boot. She would have no compunction about instructing young Faraday of his place in the social hierarchy—at least when her husband was out of earshot. Maid and valet were one and the same. They were there to serve and obey.

Meanwhile, Faraday stared in wonder as the English countryside rolled by. "I was more taken by the scenery today than by any thing else I have ever seen," he confided to his journal early in the trip. "It came upon me unexpectedly and caused a kind of revolution in my ideas respecting the nature of the earth's surface. That such a revolution was necessary is I confess not much to my credit . . ." This, a tyro's view of the "mountains" of Devonshire, a mere two days out of London and months before beholding the Alps. Faraday made no pretensions about his meager knowledge of the world, a condition he repeatedly lamented in his journal and in letters home. And that was precisely why he sat atop this carriage, being drawn ever further from the familiar and secure. As he related to his mother, "When Sir H. Davy first had the goodness to ask me whether I would go with him, I mentally said, 'No, I have a mother, I have relations here.' And I almost wished that I had been insulated and alone in London, but now I am glad that I have left some behind me on whom I can think, and whose actions and occupations I can picture in my mind."

The party sailed from Plymouth on October 17, 1813, and arrived two days later at the port of Morlaix in Brittany, the only harbor available to English vessels. If Faraday had anticipated the grandeur of civilized France, the ramshackle town of Morlaix deflated any such expectations. After being thoroughly searched by a French officer—"I could hardly help laughing at the ridiculous nature of their precautions," Faraday wrote—they slogged up the muddy lane to Morlaix's best hotel, which featured a pigsty and stables among its amenities. Faraday was obliged to wait his turn while a horse lumbered through the front door. Other than its gilded furniture, the hotel made no claim to refinement. The common areas were a menagerie of "horses pigs poultry human beings or whatever else has a connection with the house." A peek into the kitchen brought Faraday scant comfort: "I think it is impossible for an English person to eat the things that come out of this place except through ignorance or actual and oppressive hunger." Impossible or not, he ate and pronounced the hotel's cuisine "excellent and inviting."

After a four-day delay, while their carriage was reassembled and disbelieving provincial authorities verified their passports—and while Faraday caviled in his journal about French inefficiency and manners—the party was allowed to leave. A week later, Faraday was strolling the boulevards of Paris. At first, he was identified by his dress as English and was called names and spat upon. At Davy's suggestion, he purchased a suit in the French style: "I am quite out of patience with the infamous exhorbitance of these Parisians," Faraday complained about the shopkeepers. "They seem to have neither sense of honesty nor shame in their dealings. They will ask you twice the value of a thing with as much coolness as if they were going to give it to you . . ." Now clothed as a native, Faraday moved through the city like a shade among the living: "I know nothing of the language or of a single human being here, added to which the people are enemies and they are vain . . . I must exert myself to attain their language so as to join their world."

Faraday gave full attention to everything from the most mundane to the most glorious. "[E]very day presents sufficient to fill a book," he wrote Benjamin Abbott, and then proceeded to do just that. Faraday's journal records decorative elements of French apartments, specifics of an umbrella over a corn stall in the market, price fluctuations of sugar, stream patterns of Parisian fountains, wear of stair treads, the Arc de Triomphe, anatomical aspects of French pigs (which he likened in appearance to greyhounds). The journal sometimes reads like an obsessive's Baedeker, down to the dimensions of the city's great halls and monuments. By contrast, Humphry Davy's journal is haphazard and incomplete. Davy recycled a notebook which was already half-filled with experimental ideas and data. To this, he added, in seeming random fashion, poems and impressions inspired by his travels. A page of verse is succeeded by a draft report, "To attempt to decompose Nitrogen," which gives itself over to a terse "Mont Blanc Jan 5 1814 4 o'clock in the carriage."

At the Louvre with Davy, Faraday toured the vast galleries of paintings and sculpture, including famed pieces looted from Europe's greatest collections. Like Davy, he viewed the Galerie Napoléon as "both the Glory and the disgrace of France" and the country itself as "a nation of thieves." He attended French opera, which he neither understood nor appreciated. However, the grandeur of the city's churches, gardens, and public buildings did impress him.

In December, Faraday braved a downpour with thousands of others at the Tuileries Gardens to catch a glimpse of Napoléon in procession, only months before the dictator's overthrow. "He was sitting in one corner of his carriage, covered and almost hidden from sight by an immense robe of ermine, and his face overshadowed by a tremendous plume of feathers that descended from a velvet hat. The distance was too great to distinguish the features well but he seemed of a dark countenance and somewhat corpulent. No acclamations were heard where I stood and no comments."

Faraday applied the same analytical eye to French society that he did to the natural world—and, swayed by his English sensibilities, he frequently found fault. "I have found the French people in general a communicative, brisk, intelligent, and attentive set of people; but their attentions were to gain money, and for their intelligence they expect to be paid. Politeness is a general character which they well deserve; but the upper classes have carried it beyond the bounds of reason, and in politeness they lose truth and sincerity; their manners are very insinuating and kind, their address at once easy and free, and their conversation vivid and uninterrupted; but though it dwell for hours on the same subject, you can scarcely make out what the subject is: for it is certainly the most unfixed, most uninteresting, and unapplicable conversation I have met with."

"At Paris civilization has been employed mostly in the improvement and perfection of luxuries, and oftentimes in the pursuit, has neglected the means of adding to domestic and private comfort; and has even at times run counter to it. In ornaments indeed the Parisians excell, and also in their art of applying them; but in the elegance of appearance, utility is often lost . . . At Paris every thing yields to appearance, the result of what is called fine taste: the tradesman neglects his business to gain time to make appearance; the poor gent starves his inside to make his outside look well; the jeweller fashions his gold into trinkets for show and ornament; and so far does this love of appearance extend, that many starve in a garret all the week to go well dressed to the Opera on Sunday evening." Still, "as a stranger who had not always opportunities," Faraday was impressed by the ordinary citizen's—and his own—access to state museums and libraries.

Faraday reveled in his day-to-day interactions with Humphry Davy, who introduced him to the cultural and scientific institutions of Paris and tutored him in all manner of laboratory techniques and social customs. "[T]he constant presence of Sir Humphry Davy was a mine inexhaustible of knowledge and improvement," Faraday wrote a friend back home. Still, Faraday resented his role as Davy's valet. True, Davy's regular valet had backed out of the trip at the last minute, fearing for his life. And, true, Davy had promised to hire a replacement in Paris. But as weeks stretched into months, no one suitable could be found. It was not that the duties thrust upon Faraday were onerous: monitoring expenses, arranging meals and accommodations, supervising local servants. And Davy was fully capable of dressing himself. No, it was the designation itself—*valet*—that rankled. However humble his roots, Faraday was acutely aware of his station in society—and it was not that of servant. On this last point, he and Lady Davy differed, and she let Faraday know her position at every opportunity. There was a hidden cost to Faraday's Continental education: a seething outrage at his mentor's wife. For now, silence was his only retort—a silence that ripped his self-esteem and left him longing for home. Was the acquisition of knowledge truly worth such a price?

Discovery was never far from Humphry Davy's mind. The time in Paris was a chance to interact with prominent French chemists—Joseph Louis Gay-Lussac, Claude-Louis Berthollet, Louis-Nicolas Vauquelin, Charles Bernard Desormes, Michel-Eugene Chevreul—learn about their work, spark each other's creativity. Davy's portable laboratory was his weapon in some good-natured—but decidedly nationalistic—experimental jousting with his French colleagues. To this scene, Faraday was both pupil and participant.

Shortly after their arrival, a delegation led by physicist André-Marie Ampère delivered some unusual crystals derived from the ashes of seaweed (which was used in the manufacture of saltpeter for gunpowder). When heated, the crystals emitted a remarkable violet vapor, prompting Gay-Lussac to name the new substance *iode,* after the Greek for violet. French chemists had labored without success for over a year to determine its elemental makeup. Davy immediately set to work and within a few days concluded that Gay-Lussac's compound was, in fact, a wholly new element. In view of the element's chemical similarities to chlorine, Davy rechristened it *iodine* and fired off an announcement to the Royal Society. Gay-Lussac was flummoxed by the speed with which Davy had solved the problem and claimed credit for iodine's discovery. Davy, for his part, was miffed that Gay-Lussac almost immediately published studies without acknowledging their discussions. Or, as

Faraday complained to Benjamin Abbott, "The French Chemists were not aware [of] the importance of the subject until it was shewn to them, and now they are in haste to reap all the honours attached to it." In the end, science trumped petty jealousies; within a year, the zealous chemists had fully established iodine's properties. To all this, Faraday was witness. "[A]s is the practice with him," Faraday announced with sublime understatement to a friend, Davy "goes on discovering."

On December 29, 1813, after three months in Paris, the Davys, Faraday, and Mrs. Meek headed toward the south of France. As the carriage hurtled through the forest of Fontainebleau palace, Faraday's mood brightened. "I do not think I ever saw a more beautiful scene than that presented to us on the road. A thick mist which had fallen during the night, and which had scarcely cleared away had, by being frozen, dressed every visible object in a garment of wonderful airiness and delicacy." Davy was equally moved by the surreal vista. His poetic response reflects a worldliness that his protégé yet lacked:

Nearer I behold
The palace of a race of mighty kings;
But now another tenants. On these walls,
Where erst the silver lily spread her leaves—
The graceful symbol of a brilliant court—
The golden eagle shines, the bird of prey,—
Emblem of rapine and of lawless power:
Such is the fitful change of human things:
An empire rises, like a cloud in heaven,
Red in the morning sun, spreading its tints
Of golden hue along the feverish sky,
And filling the horizon;—soon its tints,
Are darken'd, and it brings the thunder storm,—
Lightning, and hail, and desolation comes;

But in destroying it dissolves, and falls
Never to rise!

In the predawn darkness of the next day, Faraday mused on his good fortune: " 'Tis a pleasant state, almost audibly to the mind, the novelty of present circumstances: that the Loire is on my right hand; that the houses to the left contain men of another country to myself; that it is French ground I am passing over, and then of the distance between myself and those rest who alone feel an interest for me, and enjoy the feeling of independence and superiority we at present possess over those sleeping around us. We seem tied to no spot, confined by no circumstances, at all hours, at all seasons and in all places. We move with freedom; our world appears extending and our existence enlarged. We seem to fly over the globe rather like satellites to it than parts of it, and mentally take possession of every spot we go over."

As they penetrated deeper into the French countryside, the pigsties of Morlaix were forgotten. In his journal, Faraday now describes accommodations as tolerably clean, food and wine good, innkeepers cheerful, inhabitants obliging. Of the French bread, he practically drools in anticipation of his next baguette: "It has a degree of positive excellence that one would not wish it surpassed; beyond what it is would be undue luxury."

After brief stops at the Pont du Gard and the amphitheater at Nîmes (where Faraday noted dimensions and Davy penned verse), the party arrived at the Mediterranean. Here they rested until mid-February 1814, when they commenced the difficult passage over the Maritime Alps toward Italy. Lady Davy and her maid rode in the carriage, while the two men ascended on foot. Faraday, having donned a second coat and a nightcap against the cold, carried a barometer to gauge the elevation. Beside him, Davy pointed out features of geological interest. The snow grew so deep that the women, luggage, and dismantled carriage were loaded onto sledges and hauled the rest of the way. Once over the mountains, they resumed travel by carriage and boat through Turin and Genoa to Florence. It was on a storm-tossed boat in the Gulf of Genoa that Lady Davy grew faint and stopped speaking. Faraday later confided to Benjamin Abbott that her rare silence was well worth the risk to their lives.

At the Accademia del Cimento in Florence, Faraday saw Egyptian mummies, life-sized anatomical models made of wax, and the telescope with which Galileo had discovered Jupiter's moons. After several days of follow-up work on iodine, Faraday witnessed Davy's remarkable experiment with diamonds. The composition of diamond was a long-standing subject of dispute: Was it formed solely of carbon, like graphite, charcoal, and "pure" soot; or was it an admixture of carbon and other elements? And if it were truly pure carbon, then how could a single element be at once the most coveted gemstone and the major constituent of the filth inside a chimney flue? Neither electricity nor fire were sufficient to rend a diamond into its constituent elements; but the prodigious heat from the concentrated rays of the sun might be. The Grand Duke of Tuscany had commissioned a "burning glass," which the Accademia del Cimento now put at Davy's disposal. Consisting of two convex lenses, 15 inches and 3 inches in diameter, separated by 3½ feet, the great glass focused sunlight onto the diamond, which was situated within a tiny glass globe containing pure oxygen. The diamond gradually evaporated over many hours, leaving behind only carbon dioxide gas. (The volatilized carbon from the diamond had united with the surrounding oxygen.) Were diamond other than pure carbon, Davy reported to the Royal Society, gases other than carbon dioxide would have been formed. Clearly, an element's material makeup is not the sole basis of its physical properties; the underlying molecular arrangement determines whether, say, carbon manifests itself as a diamond, graphite, or soot.

Upon the party's arrival in Rome in April 1814, Faraday wrote to his mother, brimming with excitement: "Tell B[enjamin Abbott] I have crossed the Alps and Appenines, I have been at the Jardin des Plantes, at the museum arranged by Buffon, at the Louvre among the *chefs-d'oeuvre* of sculpture and the masterpieces of painting, at the Luxembourg palace amongst Rubens' works, that I have seen a GLOWWORM!!! water-spouts, torpedo [i.e., a sting ray] at the Academy del Cimento as well as St Peter's and some of the antiquities here and a vast variety of things far too numerous to enumerate." But loneliness weighed heavily on Faraday: Language distanced him from the locals; situation and class from his traveling companions. "Whenever a vacant hour occurs," he told his mother, with an implicit dig at the Lady Davys of the world, "I employ it by thinking on those at home. Whenever present circumstances are disagreeable, I amuse myself by thinking on those at home. In short when sick, when cold, when tired, the thoughts of those at home are warm and refreshing balm to my heart. Let those who think such thoughts useless, vain and paltry think so still; I envy them not their more refined and more estranged feelings. Let them look about the world unencumbered by such ties and heart-strings and let them laugh at those who, guided more by nature, cherish such feelings. For me, I will still cherish them in opposition to the dictates of modern refinement as the first and greatest sweetness in the life of man."

Those back home missed Faraday as much as he missed them, especially Benjamin Abbott. In one letter, he complained to Faraday of his job as a merchant's clerk and of his own lonely efforts in science: "I have no scientific companion[s] . . . so fond of Science as to stand at a furnace till their Eyes are scorched or risk a convulsion of their Muscles from the unexpected touch of a Voltaic Battery—such a one as this would be a treasure to me, such a one as could catch my ideas and pursue them in conjunction with me and to whom half a word might suffice to convey more than a whole lecture does to some—but for such a one I must wait till *you* return . . ."

In Rome, as in Paris, Faraday drew a distinction between the city and its inhabitants. He was impressed by the great works—the Coliseum, St. Peter's, Constantine's Arch, the Trevi Fountain, the Forum—to which he applied words like *picturesque, astonishing, singular*. But the citizenry were another matter. Although he admitted that the Italians respected the English "almost to adoration," their lack of industriousness pained him. "The civilization of Italy seems to have hastened with backward steps in latter years, and at present there is found there only a degenerate idle people, making no efforts to support the glory that their ancestors left them . . . Rome is at this day not only a memento of decayed majesty, in the ruins of its ancient monuments and architecture, but also in the degeneracy of the people."

During the summer of 1814, Faraday accompanied Davy on scientific outings to Naples and Mount Vesuvius before the entire party embarked for an extended stay at a villa on the shores of Lake Geneva. Here Davy hobnobbed with a number of scientists, including chemist Gaspard de la Rive and his physicist son, Auguste, with whom Faraday maintained a decades-long correspondence. Davy's acolyte evidently made a positive impression on the Genevan scientists, one of whom commented, "We admired Davy, we loved Faraday." It was also in Geneva that Lady Davy made what might have been her most egregious affront against Faraday. Imagine Faraday's excitement when told that he and the Davys were invited to a dinner party hosted by Jane Marcet, author of the book that had awakened his interest in chemistry. Imagine, next, his horror when ordered by Lady Davy—in front of the assembled guests—to take his meal in the kitchen with the servants. He shuffled off while the others sat down at the table. At dinner's end, as the ladies were adjourning to the parlor, Jane Marcet's husband, Alexander, announced in a stage whisper, "And now, my dear Sirs, let us go and join Mr. Faraday in the kitchen."

Faraday's relationship with Lady Davy evidently reached the

breaking point upon their return to Rome in November 1814. In a despondent letter to Benjamin Abbott, Faraday practically screams, "Alas! how foolish perhaps was I to leave home, to leave those whom I loved and who loved me . . . And what are the boasted advantages to be gained? Knowledge. Yes, knowledge, but what knowledge? Knowledge of the world of men, of manners, of books and of languages—things in themselves valuable above all price, but which every day shews me prostituted to the basest purposes. Alas! how degrading it is to be learned when it places us on a level with rogues and scoundrels! How disgusting when it serves but to shew us the artifices and deceits of all around! How can it be compared with the virtue and integrity of those who, taught by nature alone, pass through life contented, happy, their honor unsullied, their minds uncontaminated, their thoughts virtuous—ever striving to do good, shunning evil and doing to others as they would be done by? . . . What a result is obtained from knowledge and how much must the virtuous human mind be humiliated in considering its own powers, when at the same time they give him such a despicable view of his fellow creatures! Ah, Ben, I am not sure that I have acted wisely in leaving a pure and certain enjoyment for such a pursuit."

Alarmed, Abbott wrote back immediately, asking what had triggered such a rush of emotion. At last, Faraday revealed his true feelings about Lady Davy: "She is haughty and proud to an excessive degree and delights in making her inferiors feel her power. She wishes to roll in the full tide of pleasures such as she is capable of enjoying but when she can with impunity—that is, when her equals do not notice it and Sir H. is ignorant of it—she will exert herself very considerably to deprive her family of enjoyments. When I first left England, unused as I was to high life and politeness, unversed as I was in the art of expressing sentiments, I [came] . . . under the power to some degree of one whose whole life consists of forms, etiquette, and manners. I believed at that time that she hated me and her evil disposition made her endeavour to thwart me in all my views and to debase me in my occupations."

As 1814 drew to a close, a more confident and polished Faraday had learned to fend off Lady Davy's barbs. "At present I laugh at her whims which now seldom extend to me," he assured Abbott. Still, their arguments left Humphry Davy in a precarious spot and sometimes led to an awkward coolness among them all for days at a time. (About the marriage, Humphry Davy's brother, John, suggested frankly that there "was an oversight, if not a delusion, as to the fitness of their union; and it might have been better for both if they had never met.")

Despite the ongoing tensions, Faraday dove into the "licentiousness and riot" of Rome's Winter Carnival, when "the Pope loses his power for a week or more." To Benjamin Abbott, he wrote, "Rome glittered with Princes, Princesses, Dukes, Lords, Spaniards, Italians, Turks, Fools &c., all of which were in profusion." The clatter of hooves practically rises from the pages when Faraday describes the furious running of the horses in the Corso. He also frequented the predawn masquerade balls, dressed either in a nightgown and nightcap or in a priest's robe and hood. (After one ball, he fell in line with what he took to be other "priestly" revelers; it was not until he saw the body that he realized it was a funeral procession.)

By early 1815, with the Carnival's raucous excitement behind him, Faraday wearied of his peripatetic existence. The free-flying "satellite" who had soared over the French countryside just a year earlier had fallen to earth. "As for me," Faraday wrote his friend Huxtable, "like a poor unmanned, unguided skiff, I pass over the world as the various and ever-changing winds may blow for me, for a few weeks I am here, for a few months there, and sometimes I am I know not where, and at other times I know as little where I shall be." Faraday's "ever-changing winds"—the Davys—had intended to embark next for Greece and Turkey. But plague had broken out and Italian troops were mustering in the wake of Napoléon's escape from Elba. In March 1815, Faraday received welcome news from Humphry Davy: They were going home. The party left Italy and traveled in haste through Germany and Holland to Brussels, where they awaited passage on a ship sailing from Ostend. From Brussels, Faraday informed his mother of their unexpected change in plans: "It is with no small pleasure I write you my last letter from a foreign country, and I hope it will be with as much pleasure you will hear I am within three days of England . . . I have a thousand times endeavoured to fancy a meeting with you and my relations and my friends, and I am sure I have as often failed: the reality must be a pleasure not to be imagined nor to be described . . . I come home almost like the prodigal, for I shall want everything." He closed with the particulars of his arrival in England and signed the letter. Then he appended a postscript: " 'Tis the shortest and (to me) the sweetest letter I ever wrote you."

4

FEAR AND CONFIDENCE

Man's greatest asset is the unsettled mind.

—ISAAC ASIMOV

The Faraday who returned from the Continent in April 1815 was wholly recast from the artless young man who had embarked nineteen months earlier. He had strolled the boulevards of the great cities of Europe; cast his eyes upon sublime works of art; touched the foundation stones of Western civilization; engaged with a host of humanity, from anonymous peasants to noted scientists; studied at the side of a brilliant mentor; and withstood blistering assaults on his self-esteem. In the acquisition of knowledge, he had not merely separated kernels from husks, in the down-to-earth metaphor of Isaac Watts; he had extracted—and now possessed—gold from dross. Faraday's lack of schooling, his coarseness, and his youthful inexperience had been mitigated by his journey. He had stood toe to toe with aristocrats and had emerged with an undiminished sense of self; he harbored no desire to join the gentry in their meaningless game of manners. Having traversed a wider geographic and social world than he had ever encountered—perhaps even imagined—Faraday realized that his lofty aspirations to become a scientist might truly be achievable. Before, thanks to Humphry Davy, he had planted a foot in the closing door of opportunity. Now, again thanks to Davy, the door swung open wide.

Faraday resumed his work at the Royal Institution, with a modest pay raise, two rooms instead of one, and—at Davy's insistence—a new title: Assistant and Superintendent of the Apparatus and Mineralogical Collection. He was much more a junior scientist than the "fag and scrub" he used to be. While Davy was away, lecturing, fishing, and socializing, Faraday performed chemical analyses of substances sent back from the road, such as pieces of ore or samples of air from a coal mine. He was invited by Davy to Italy to assist in the unrolling of papyri at Herculaneum, but was too ensconced at the Royal Institution to leave. Or perhaps he wished to keep his distance from Lady Davy. (By this time, his mentor's married life had become famously tumultuous, as related by Sir Walter Scott, Jane Davy's cousin: "She has a temper and Davy has a temper . . . and they quarrel like cat and dog, which may be good for stirring up the stagnation of domestic life, but they let the world see it, and that is not so well.")

Davy regularly lauded Faraday in research papers for his "accuracy and steadiness of manipulation." In one letter to Faraday questioning the purported discovery of a new acid by Irish chemist Michael Donovan, Davy writes, "Pray make an investigation of this subject. I think you are a better chemist than Donovan." From Rome, in October 1818, he tells Faraday of a communication he received from Charles Hatchett, a chemist for whom Faraday had conducted an analysis: "Mr. Hatchett's letter contained praises of you which were very gratifying to me & pray believe me there is no one more interested in your success & welfare than your sincere well wisher & friend, H. Davy." And this, seven months later, from Florence: "It gives me great pleasure to hear that you are comfortable at the Royal Institution & trust you will not only do something good

& honorable for yourself; but likewise for Science." Faraday had become Davy's trusted hands in the laboratory, ensuring the absent
professor's productivity back home while he and his wife pursued
their peripatetic lifestyle. And while the great man's plaudits were
nice, Faraday must have been pondering his own future. When
would Davy release him to become an independent researcher?

When not doing Davy's bidding, Faraday played a crucial role
assisting his on-site superior, Professor William T. Brande, in both
the lecture hall and laboratory; compiling the Miscellanea section
of the Royal Institution's *Quarterly Journal of Science,* and conducting
commercial chemical analyses. Only three years older than Faraday, Brande was as pedestrian as Davy was flamboyant. Nevertheless, he was a steadfast researcher from whom Faraday could learn.
One acquaintance remarked, with an obvious nod toward Davy,
that Brande "was never brilliant or eloquent but his experiments
never failed." That Brande, like Davy, recognized Faraday's experimental finesse is evident in the Preface to Brande's *Manual of Chemistry:* "I have uniformly received the active and able assistance of Mr
M. Faraday, whose accuracy and skill as an operator have proved
an essential service in all my proceedings." Already by September
1815, the handwriting in the official Laboratory Notebook changes
"from the large running letters of Brande to the small, neat characters of Faraday."

In the autumn of 1815, Faraday assisted Davy in a project of immense importance: the development of the miner's safety lamp.
Several years earlier, an explosion inside the Felling Colliery in
northern England had taken the lives of ninety-two men and
plunged their families into poverty. This was only the latest in a
string of similar disasters that had laid a stain on Britain's rapid industrialization and its attendant hunger for coal. At the time, the
only way to effectively illuminate a mining tunnel was with an open
flame. And the only way to detect the buildup of flammable gases
was to have a "fire-man," cocooned in damp cloth, inch ahead of

the miners with a candle stuck on the end of a long pole. Distressed by the loss of life and the heart-rending accounts of the widows and orphans, a committee of physicians, clergymen, and landowners asked Humphry Davy to attack the problem scientifically. Davy obtained gas samples from a number of mines and, with Faraday, quickly deduced that accumulated methane was responsible for the explosions. By combining methane with various quantities of oxygen, then igniting the mixture with a flame, the two scientists determined that methane explodes only in the presence of sufficient oxygen. Davy's key innovation was to surround the lamp flame with a cylindrical metal mesh that absorbed oxygen from the adjacent air and, thus, deprived the methane gas of its trigger. Not only did the lamp provide safer illumination, but the varying glow of the red-hot mesh indicated the amount of methane in the tunnel.

Upon testing Davy's lamp, the chief engineer of the Wall's End Colliery exulted, "We have subdued this monster." Davy himself descended into the mines to instruct all in the proper use of his lamp. The days of the candle on a pole were over. To the captains of British industry, and surely to Faraday, here was a powerful demonstration of the potential of science to solve practical problems.

Davy did not patent his safety lamp. His sole compensation was a set of commemorative dishware from the grateful mine owners, so that he might reflect upon his extraordinary contribution even while he ate. As to claims by various inventors that Davy had appropriated their work, Faraday himself answered the accusations in a lecture at the City Philosophical Society in 1817, and again more than a decade later at the Royal Institution: "I was witness in our laboratory to the gradual and beautiful development of the train of thought and the experiments which produced it. The honour is Sir H. Davy's, and I do not think that this beautiful gem in the rich crown of fame which belongs to him will ever again be sullied with the unworthy breath of suspicion." (The latter defense was all the more generous considering that, when Faraday was similarly charged on an unrelated project, Humphry Davy would become one of his most vocal accusers.)

Both Faraday and, more reluctantly, Davy admitted that the safety lamp could trigger an explosion in the presence of strong drafts or too much suspended coal dust. Indeed, the number of mine fatalities remained steady in the years after the lamp's introduction, although this may be due in part to an increased amount of mining or the improper use of the lamp. When an exhibit opened at the Royal Institution in 1826 of an *improved* safety lamp, Faraday penciled in mischievously: "The opinion of the inventor."

In 1816, Davy asked Faraday to analyze a sample of caustic lime, a corrosive calcium compound, sent to him from Tuscany. This became Faraday's first published paper. It was a time, Faraday later wrote, "when my fear was greater than my confidence, and both far greater than my knowledge." In subsequent projects, Faraday disproved a purported discovery of a new element, sirium; investigated the chemical properties of various compounds of carbon, mercury, iron, and zinc; and collaborated with surgical instrument maker James Stodart in pioneering research on alloys of steel. While not all of his experiments bore fruit, he tried always to maintain a high degree of care and rigor in their conduct. But on one occasion, he erred. In 1819, Davy published a study of the compounds of phosphorus based in part on Faraday's laboratory work. The paper took issue with the findings of the noted Swedish chemist Jöns Jacob Berzelius. In response, Berzelius fired this broadside in the premier chemistry journal of the day, *Annales de chimie et de physique:* "If M. Davy would be so kind as to take the pains of repeating these experiments himself he should be convinced of the fact that when it comes to exact analyses, one should never entrust them into the care of another person; and this is above all a necessary rule to observe when it comes to refuting the works of other chemists who have not shown themselves ignorant of the art of making exact experiments." The public drubbing by

Berzelius was an excruciating lesson to Faraday, who never again allowed his data to be published before carefully verifying the results. Although there is no mention of the incident in his writings, Faraday's thoughts might well have been reflected in a letter he wrote many years later on behalf of a colleague who had published an erroneous assertion: "[W]e are all liable to error, but . . . we love the truth, and speak only what at the time we think to be the truth; and ought not to take offence when proved to be in error, since the error is not intentional, but be a little humbled, and so turn the correction to good account." While others subsequently found fault with Faraday's reasoning or his conclusions, his laboratory findings became unimpeachable.

Faraday's growing reputation as an analytical chemist brought him lucrative business and government contracts. For a fee, he analyzed water, wines, paper, rust, and compounds with such tongue-twisting names as Dutch turf ash, Paligenetic tincture, and *Baphe eugenes chruson* (a golden dye). At the Admiralty's request, he tested methods of drying meats and fish, measured gases emitted by aging eggs, and assessed the purity of military-issue oatmeal. In 1820, Faraday testified in court on behalf of an insurance company regarding the use of a flammable oil in the refining of sugar. The refiner sought payment from its insurer following a fire in its factory; the insurer countered that the refiner's use of the oil voided the policy. Faraday was dismayed to learn that the refiner had hired its own expert witnesses: Davy and Brande. In the end, Faraday's scientific testimony was the more compelling; he had conducted experiments on the oil in question while his senior colleagues had relied heavily on models and inference. Nevertheless, the court directed the insurer to pay, since no fraud had been intended by the refiner.

Something of Faraday's frenzied worklife—and also his supreme humility—is glimpsed in an apologetic letter he wrote in 1818 to a surgeon who had asked his (free) advice about lightning rods: "I am

Faraday's basement laboratory at the Royal Institution.

continually saying to myself that I have not yet time to do this or that thing, and yet, when the performance has been delayed until an hour rendered inconvenient from its lateness, when it must be done, I have suspected that an undue admission of small but dangerous delays has been the cause of the whole evil. I have not written to you before, because at each time when I thought of it I had something else in hand; and yet I must confess that many convenient opportunities for the purpose have passed away since I received yours. I hope you will not deny me my pardon. My honest confession ought to mediate for me in some degree; and though a promise not to do so again will not remove the error already committed, it may perhaps tend to diminish the punishment not yet inflicted." In suggesting that the surgeon run his own lightning-conducting wire down the chimney, Faraday wryly cautions, ". . . only take care you do not kill yourself, or knock down the house."

Meanwhile, Faraday maintained his efforts at self-improvement, delving into the ready resources of the Royal Institution's library and attending its evening classes in oratory. He continued to correspond with Benjamin Abbott so his writing might become more spontaneous: "I would, if possible, imitate a tree in its progression from roots to a trunk, to branches, twigs & leaves, where every alteration is made with so much ease & yet effect that, though the manner is constantly varied, the effect is precise and determined." And despite the heavy demands of his job, he rushed to weekly meetings of the City Philosophical Society. Only now, unlike most CPS members, Faraday was a practicing scientist. So highly was he regarded by his peers that one of them composed an anonymous paean to him that reads, in part:

His powers, unshackled, range from pole to pole;
His mind from error free, from guilt his soul.
Warmth in his heart, good humor in his face,

A friend to mirth, but foe to vile grimace;
A temper candid, manners unassuming,
Always correct, yet always unpresuming.
Such was the youth, the chief of all the band;
His name well known,
    Sir Humphry's right hand.
With manly ease toward the chair he bends,
With Watts's Logic at his finger-ends.

On January 17, 1816, Faraday delivered a lecture to the CPS titled "On the General Properties of Matter." It was the first in a series of seventeen lectures Faraday delivered between 1816 and 1818 that encompassed the whole of inorganic chemistry. In his fifth lecture, he suspended the didactic narrative to declare his core scientific philosophy: "The [natural] philosopher should be a man willing to listen to every suggestion, but determined to judge for himself. He should not be biassed by appearances; have no favourite hypothesis; be of no school; and in doctrine have no master. He should not be a respecter of persons, but of things. Truth should be his primary object. If to these qualities be added industry, he may indeed hope to walk within the veil of the temple of nature."

In an essay for a CPS study group, Faraday tackles an aspect of the scientist that would become one of his hallmark strengths: the synergy between imagination and judgment. "[The satisfaction] resulting from judgment," he writes, "is the noblest and that of imagination the most enticing but where a union of the two takes place in a strong degree there will always be a great cause for our admiration. The fairest fruits are the highest hung and few there are who can reach them." In a follow-up essay, he rhapsodizes about nature's beauty—at least until his left-brain sensibilities kick in: "A sensitive mind will always acknowledge the pleasures it receives from a luxuriant prospect of nature; the beautiful mingling and gradations of colour, the delicate perspective, the ravishing effect of light and shade, and the fascinating variety and grace of the outline, must be seen to be felt; for expressions can never convey the extatic joy they give to the imagination, or the benevolent feeling they create in the mind. There is no boundary, there is no restraint 'til reason draws the rein, and then imagination retires into its recesses, and delivers herself up to the guidance of that superior power."

Although essentially uninterested in politics, Faraday revealed his political inclinations in a series of anagrams in his *Common Place Book:*

Democratical–comical trade
Revolution–to love ruin
Old England–golden land
Radical Reform–Rare Mad Frolic
Universal Suffrage–guess a fearful ruin
Monarch–march on

With professional and family obligations saturating his calendar, Faraday had little time to socialize–and even less to devote to the notion of romance. Nature was his sole mistress, with whom he might consort on his own terms and fulfill the longings of his soul. The idea of falling in love–much less pursuing it–was as ridiculous as it was antithetical to his progress as a scientist. His aversion to stirrings of the heart was made plain in his *Common Place Book* around 1817:

What is the pest and plague of human life?
And what the curse that often brings a wife?
    'tis Love.

What is the power that ruins man['s] firmest mind?
What that deceives its host when alas too kind?
What is it that comes in false deceitful guise

Making dull fools of those that before were wise?
    'tis Love.

Whether brought on by Humphry Davy's stormy marriage or perhaps a friend's romantic travails, Faraday's huffy anthem to bachelorhood exhausts itself for another twenty-five lines before giving Love the heave-ho:

Love, then thou 'ast nothing here to do.
Depart, Depart to *yonder* crew.

Of course, Love dances to an unpredictable rhythm. Faraday threw aside his misogynist credo when he met nineteen-year-old Sarah Barnard, a fellow congregant in his Sandemanian church and sister of one of his CPS colleagues. Sarah's brother had read her Faraday's screed against love, so it must have startled her when the up-and-coming chemist laid claim to her heart. In the matter of courtship, Faraday was no Humphry Davy, who could set a torrid pen to paper and unleash the ardor of ages. At first, Faraday clothed his passion in a veneer of rationality—the suitor as scientist. When that failed to move Sarah, he let fly with more emotion-packed pleas, becoming fully the heartsick fool he had once mocked in verse. "You know me as well or better than I do myself," he wrote to Sarah in 1820. "You know my former prejudices, and my present thoughts—you know my weaknesses, my vanity, my whole mind; you have converted me from one erroneous way, let me hope you will attempt to correct what others are wrong . . . Again and again I attempt to say what I feel, but I cannot. Let me, however, claim not to be the selfish being that wishes to bend your affections for his own sake only. In whatever way I can best minister to your happiness, either by assiduity or by absence, it shall be done. Do not injure me by withdrawing your friendship, or punish me for aiming to be more than a friend by making me less; and if you cannot grant me more, leave me what I possess, but hear me."

Sarah showed the letter to her parents, who promptly shipped her off to Ramsgate, on England's southeast coast. Here, they felt, in the company of her elder sister, she might weigh her options in peace. Disconsolate, Faraday followed. At first, he made little headway in his plight. "I wished for a moment," he confided to his journal, "that memory and sensation would leave me, and that I could pass away into nothing." Several days later, he and Sarah toured the cliffs of Dover, and here, amidst the magnificence, he received the answer he was seeking.

Faraday's journal entry for the day so resounds with light, air, and joy that one almost tastes the salty breeze:

> The cliffs rose like mountains . . . to an immense height with summits and ridges towering in the air . . . At the foot of these cliffs was the brilliant sparkling ocean, stirred with life by a fresh and refreshing wind, and illuminated by a sun which made the waters themselves seem inflamed. On its surface floated boats, packets, vessels, beating the white waves, and making their way against the feigned opposition of the waters. To our left lay Dover, with its harbour and shipping equally sheltered and threatened by the surrounding hills; and opposite were the white cliffs of the French coast . . . The whole was beautiful, or magnificent, as the mind received its tone from successive thoughts, and almost became sacred when the eye wandered towards the arch cliff, for there Shakespeare's spirit might be fancied sitting on the very verge, absorbed in contemplation of its grandeur . . .
>
> I can never forget this day. Though I had ventured to plan it, I had little hope of succeeding. But when the day came, from the first waking moment in it to the last it was full of interest to me: every

circumstance bore so strongly on my hopes and fears that I seemed
to live with thrice the energy I had ever done before . . .

But now that the day was drawing to a close, my memory recalled
the incidents in it, and the happiness I had enjoyed; and then my
thoughts saddened and fell, from the fear I should never enjoy such
happiness again . . .

I could not master my feelings or prevent them from sinking, and I
actually at last shamed myself by moist eyes . . . It is certainly strange
that the sincerity and strength of affection should disable me from
judging correctly and confidently of the heart I wish to gain, and
adopting the best means to secure it . . . But sincerity takes away all
the policies of love. The man who can manage his affairs with the
care and coolness of his usual habits is not much in earnest. Though
the one who feels is less able than the one who does not to take advantage of circumstances as they occur, still I would not change the ho-
nourable consciousness of earnest affection and sincerity for the cool
caution and procedure of the mind at ease, though the first were
doomed to failure and the last were blessed with success.

Returning to London, Faraday wrote often to Sarah, even when
there was evidently nothing to say:

My dear Sarah,—It is astonishing how much the state of the
body influences the powers of the mind. I have been thinking
all morning of the very delightful and interesting letter I
would send you this evening, and now I am so tired, and yet
have so much to do, that my thoughts are quite giddy, and
run round your image without any power of themselves to
stop and admire it. I want to say a thousand kind and, believe
me, heartfelt things to you, but am not master of words for

Pencil sketch of Faraday's wife, Sarah Barnard.

the purpose; and still, as I ponder and think on you, chlorides, trials, oil, Davy, steel, miscellanea, mercury, and fifty other professional fancies swim before and drive me further and further into the quandary of stupidness.

Sarah Barnard and Michael Faraday were married without fanfare on June 12, 1821. He had instructed Sarah's sister, "There will be no bustle, no noise, no hurry occasioned even in one day's proceeding. In externals, the day will pass like all others, for it is in the heart that we expect and look for pleasure." The Faradays moved into a second-floor apartment in the Royal Institution, where they lived for the next forty-one years. Sarah did not share her husband's scientific interests; instead, noting the frenetic pace at which he drove himself, she was "quite content to be the pillow of his mind." After spending hours in the laboratory, trying to unravel nature's intricate plan piece by piece, he could ascend the stairs to a space where there was, at least, no human mystery to decipher. With her honesty of spirit and lack of pretension, Sarah completed Faraday's personal universe. And he loved her for it.

Faraday wrote to Sarah in 1822 while she was away visiting relatives: "I am tired of the dull detail of things, and want to talk of love to you . . . The theme was a delightful and cheerful one before we were married, but it is doubly so now. I now can speak not of my own heart only, but of both our hearts. I now speak, not only with any doubt of the state of your thoughts, but with the fullest conviction that they answer to my own. All that I can now say warm and animated to you, I know that you would say to me again. The excess of pleasure which I feel in knowing you mine is doubled by the consciousness that you feel equal joy in knowing me yours. Oh, my dear Sarah, poets may strive to describe and artists to delineate the happiness which is felt by two hearts truly and mutually loving each other; but it is beyond their efforts, and beyond the thoughts and conceptions of anyone who has not felt it. I have felt it and do feel it, but neither I nor any other man can describe it; nor is it necessary. We are happy, and our God has blessed us with a thousand causes why we should be so."

To Sarah's surprise, a month after they were wed, Faraday stood up before the elders and congregants in the Sandemanian chapel and voiced his formal declaration of faith. When she asked him later why he had not told her beforehand, he remarked, "That is between me and my God." Although now an "official" member of the church, and thus obligated to accept a literal reading of the Bible, Faraday resolutely kept his religion from intruding upon his science. Faith guided his conduct; reason and experiment guided his work. The philosophical and behavioral strictures of the Sandemanians served to channel Faraday's abundant energy toward his scientific pursuits. In particular, the Sandemanian belief in human fallibility permitted him to fearlessly explore frontiers of science, with full knowledge that some of his conjectures would be overturned. There was no purpose in hitching his ego to the correctness of his conclusions. Faraday noted Job's admonition in his personal Bible: "If I justify myself, mine own mouth shall condemn me; if I say, I am perfect, it shall also prove me perverse" (Job 9: 20).

As 1821 edged into autumn, Michael Faraday found himself in an untenable spot. Now thirty years old, married, and settled in his faith, he had already garnered a measure of acclaim for his work. Yet in many ways, he was still Humphry Davy's assistant—a glorified valet, really—one day extolled for his experimental acumen, and the next, asked to send his boss a packet of dead flies for a fishing lure. Although it pained him, Faraday complied with all of Davy's requests, for Davy had been the grantor of his dreams. Without Davy, he would be cutting, pasting, and pounding paper in a daily cycle of despair. There would have been no Continental tour, no Royal Institution, no master lessons in the art of experiment. But as much as Davy respected Faraday's abilities, he was unwilling to view his able protégé as more than an adjunct to his own illustrious career. He was unable to set Faraday free.

Faraday had few local colleagues to whom he might confide his frustrations. There were simply none of sufficient standing who would intercede on his behalf, especially now that Davy had ascended to the very pinnacle of England's tightly knit scientific establishment—the presidency of the Royal Society. A cruel irony: Having once escaped servitude to a bookbinder, Faraday now faced the same predicament in the sciences.

But on September 3, 1821, Michael Faraday's vision momentarily overleaped that of some of the world's most prominent scientists. He made a discovery that would begin to unravel the fetters that bound him to Humphry Davy. And would nearly destroy his career.

5

RISING TO THE LIGHT

The most exciting phrase to hear in science, the one that heralds new discoveries, is not "Eureka!" (I found it!) but "That's funny . . ."

—Isaac Asimov

On October 1, 1820, Humphry Davy swept into the laboratory of the Royal Institution with remarkable news for Michael Faraday. While performing a demonstration before a science class, Danish physicist Hans Christian Oersted had noticed that an electrical current flowing in a wire moved a nearby magnetic compass needle. Whenever Oersted brought the compass toward the wire, something wrested the needle from its tenuous alignment with the earth's magnetic field and swung it in a different direction. Evidently, current in a wire creates its own halo of force—later proved to be a magnetic field, not from an ordinary magnet, but from an electrical impostor. Oersted's observation confirmed what some scientists had suspected: Electricity and magnetism were fundamentally related. (This hunch was based on a philosophical stance that all forces are manifestations of a single fundamental force; scientists today are still trying to prove such a "grand unified theory.")

That no one before Oersted had observed the magnetic aspect of electricity may seem astonishing in retrospect, especially when battery-powered electric circuits were common in 1820s-era laboratories, and compasses had been around for centuries. True, the influence of a current-carrying wire on a compass needle can be subtle. (I've tried. It helps to wrap the wire several times around the compass to concentrate the magnetic effect.) But, more important, most scientists at the time had been educated (indoctrinated?) to believe that electricity and magnetism were distinct phenomena. In France, for example, where the ideas of the influential eighteenth-century physicist Charles Coulomb dominated the scientific community, electricity and magnetism were understood to be different fluids that do not interact with each other. After Oersted's announcement, physicist André-Marie Ampère lamented to a friend, "You are quite right to say that it is inconceivable that for twenty years no one tried the action of the voltaic pile on a magnet. I believe, however, that I can assign a cause for this; it lies in Coulomb's hypothesis on the nature of magnetic action; this hypothesis was believed as though it were a fact [and] it rejected any idea of action between electricity and the so-called magnetic wires. This prohibition was such that when [physicist] M. Arago spoke of these new phenomena at the Institute, they were rejected . . . Every one decided that they were impossible."

The most remarkable aspect of Oersted's electromagnetic force was its unprecedented geometry. Unlike gravity or electrical force, both of which act along the beeline between objects, the new force was arrayed in circular fashion around the length of the wire. Were you to grasp the wire in your fist, your fingers would curl in the manner of the magnetic force. At every point, the magnetic force would nudge a drifting, magnetized fleck, not away from or toward the wire, but to the side. Nowhere in the vast Newtonian paradigm of mechanics had there ever been a force like this. Instead of a simple attraction and repulsion between objects—the tug of the sun on the earth or the "static cling" of lint to a sweater—here was a force that impelled objects transversely.

Like many of their scientific brethren, Davy and Faraday immediately re-created Oersted's experiment in the laboratory, minutely observing various configurations of magnetized needles and current-carrying wires. (It's no surprise that chemists would investigate a physics phenomenon: The boundary between scientific disciplines was fluid back then.) In the succeeding months, Davy pursued his electromagnetic studies alongside William Hyde Wollaston, a wealthy polymath who had made essential discoveries in physiology, optics, and chemistry. Faraday took no offense at being excluded; he was plenty busy with other matters, including his courtship of Sarah Barnard.

To explain the forces surrounding a live wire, Wollaston posited that electric current moved in spiral fashion along the wire. He predicted that a current-carrying wire would therefore spin about its longitudinal axis when a strong magnet was placed nearby, like a rolling log. In April 1821, he and Davy tried, without success, to induce such an electromagnetic rotation. Faraday arrived in the lab later that day, while the vanquished scientists were discussing the failed experiment. Nowhere in Faraday's laboratory diary for this period is there any mention of the conversation or of electromagnetic rotation.

In the summer of 1821, Faraday's close friend Richard Phillips asked him to write a review of what was currently known about electromagnetism for the journal *Annals of Philosophy*. Characteristically, before committing a word to print, Faraday read practically everything there was to read on the subject and duplicated in the Royal Institution laboratory many of the published experiments by Ampère, Arago, Oersted, and others. By summer's end, he was well-versed in the experimental and theoretical aspects of electromagnetism, if not so well in the subject's mathematically dense underpinnings.

In France, Faraday learned, Ampère had already moved far beyond Oersted's original observations and incomplete theory. While Oersted viewed electromagnetism as a "one-way" effect of an electric current on a magnet, Ampère considered its mutual aspect: If current exerts a force on a magnet, then shouldn't a magnet act likewise on a current? To prove his assertion, Ampère suspended an electrified coil from a pivot, so that the coil was free to turn. When he brought a magnet nearby, the coil aligned itself to the magnet. And if very carefully suspended, the coil even swung in response to the earth's much weaker magnetic field. The coil, Ampère noted, behaved as though it were a magnetized compass needle. In a bold theoretical leap, Ampère asserted that *all* magnetism was but electricity: Within a bar magnet must be some form of internal electric currents that together generate the external magnetic effects. Even the earth's magnetic field must arise from some vast tide of electricity within it. In Ampère's words, "All the phenomena presented by the mutual action of an electric current and a magnet discovered by M. Oersted . . . are covered by the law of attraction and of repulsion of two electric currents that has just been enunciated, if one admits that a magnet is only a collection of [internal] electric currents . . ." (It would be another century before these hypothesized internal currents were linked to electrons in the atom. Whirring around its central atomic nucleus, each electron behaves like a miniature electric current, which generates a tiny magnetic field, much as Ampère would have pictured it. However, in ordinary materials, atoms are oriented randomly and these magnetic effects cancel each other out. But electrons exhibit an additional property akin to spin that also gives rise to magnetism. In "ferromagnetic" metals, principally iron, a complex mechanism aligns the electron "spins" and endows the host object with an overall magnetism. Fundamentally, Ampère had it right.)

Ampère next posited that, if a current-carrying wire behaves like a magnet, then two adjacent wires would attract or repel each

other—depending on whether the currents are aligned or opposed—
as would a pair of magnets. This he confirmed by experiment. So
did Faraday. However, where Ampère saw a straight-line, classic
Newtonian attraction or repulsion between currents in the two
wires, Faraday saw a more complex situation. In Faraday's view,
each current generates a circular magnetic force in the space around
the wire; it is these "secondary" forces, he claimed, not any pur-
portedly exerted by the currents themselves, that interact to create
the push or pull on the wires. (The resolution of such seemingly
picky details often leads to fresh physical insights and scientific
breakthroughs—as it would in this case.) But Ampère's powerhouse
contribution was his all-encompassing theory of electromagnetism,
whose mathematical formulae elegantly—and quantitatively—
express the various interactions between electric currents and magnets. For this, noted physicist James Clerk Maxwell dubbed
Ampère the "Newton of electricity."

Faraday could not fathom Ampère's mathematical formulation
of electromagnetism. But he understood enough to convince himself that Ampère had not sufficiently made the case for the purported internal currents. Ampère claimed that these "molecular
currents" arise around particles of matter, and posed a complex
physical theory involving the flow, counterflow, and vibrations of
imponderable electrical fluids. Predictably, Faraday saw no experimental evidence for any of this speculation. Nor would he embrace
the widely held belief that electricity was a material fluid, as opposed to an energetic condition induced within a wire. Thus, while
Ampère's equations produced results that neatly matched what was
observed in nature, Faraday could not accept the new theory until
its underlying physical basis was established. "I have really been
ashamed sometimes," Faraday admitted to Gaspard de la Rive, "of
my difficulty in receiving evidence urged forward in support of
opinions on electro magnetism but when I confess my want of mathematical knowledge and see mathematicians themselves differing

about the validity of arguments used it will serve as my apology for waiting for experiment."

Of scientific ("philosophical") proof, Faraday had already declared his rigorous standards in an 1819 lecture to the City Philosophical Society:

> Nothing is more difficult and requires more care than philosophical deduction, nor is there anything more adverse to its accuracy than fixidity of opinion. The man who is certain he is right is almost sure to be wrong; and he has the additional misfortune of inevitably remaining so. All our theories are fixed upon uncertain data, and all of them want alteration and support. Ever since the world began opinion has changed with the progress of things, and it is something more than absurd to suppose that we have a certain claim to perfection; or that we are in possession of the acme of intellectuality which has or can result from human thought. Why our successors should not displace us in our opinions, as well as in our persons, it is difficult to say; it ever has been so, and from an analogy would be supposed to continue so. And yet with all the practical evidence of the fallibility of our opinions, all—and none more than philosophers—are ready to assert the real truth of their opinions . . . All I wish to point out is . . . the necessity of cautious and slow decision on philosophical points, the care with which evidence ought to be admitted, and the continual guard against philosophical prejudices which should be preserved in the mind. The man who wishes to advance in knowledge should never of himself fix obstacles in the way.

Faraday was frank with his criticisms when writing to Ampère, whom he had met on his Continental trip with Davy. "I am naturally sceptical in the matter of theories," he wrote to Ampère in 1822, "and therefore you must not be angry with me for not admitting the one you have advanced immediately. Its ingenuity and applications are astonishing and exact but I cannot comprehend how the currents are produced and particularly if they be supposed to exist round each atom or particle and I wait for further proofs of their existence before I finally admit them." Ampère's responses were always cordial; he readily acknowledged that Faraday's objections sometimes motivated further experiments. But it would not do for a lowly laboratory assistant to *publicly* criticize the work of some of Europe's most distinguished researchers. So Faraday directed that his published review of electromagnetism appear under the pseudonym "M."

In September, with his review nearly complete, Faraday decided to subject Oersted's discovery to more rigorous examination. In consequence, he came to understand that Wollaston had grasped only part of the story in predicting that an electrified wire would *spin* in the presence of a magnet. For if the wire were somehow unconstrained, Faraday now realized, Oersted's circular force would cause it to *revolve* about the pole of a fixed magnet like a tiny satellite. Or if the wire were tied down and the magnet freed, the magnet would circle the wire.

From Faraday's perspective, there was only one way to convince himself—and skeptical colleagues—of the inferred circular movement. Build a device that showed it. But how? It was easy to *imagine*, say, a featherweight compass needle—miraculously freed of gravity's bond—whirling satellite-fashion about a wire. So too with a current-carrying segment of wire in "orbit" around a substantial magnet. But crossing over to the laboratory realm, the predicted circular motion invariably brought about a tangle of physical impossibilities: wires passing through magnets and magnets through wires; electricity coursing through broken circuits; gravity loosening its ubiquitous grasp. Surely a number of Faraday's contemporaries had already conjured electromagnetic rotations in their minds. Yet none of them had succeeded in assembling a real-world version of such a device.

On September 3, 1821, with his wife Sarah's fourteen-year-old brother George Barnard at his side, Faraday cobbled together a crude apparatus from materials in the lab. First, he secured a bar magnet upright in a cup of mercury, leaving only the top part of the magnet exposed. (Mercury, a liquid metal, conducts electricity; thus the space around the immersed magnet was a metal "sea," able to conduct a battery's current.) From an electrical terminal above, Faraday suspended a short length of wire, buoyed by a cork, so that the wire's lower end dipped into the mercury. The dangling wire and mercury now formed an unimpeded path for electrical flow. And even if the wire were to move within the mercury—as Faraday hoped—the circuit would remain unbroken. He and George looked on expectantly as power was applied from a battery. The wire started to whirl in a conical path around the magnet. Faraday is said to have shouted, "There they go! There they go! We have succeeded at last!"

The flexible, current-carrying wire "felt" the force of the upright magnet, nudging it around in a conical path. The analogous effect occurred when Faraday hung a slender magnet from above and affixed a vertical wire within the mercury-filled cup: The magnet circled the upright electrified wire. Faraday's rough-hewn experiment marked the first time that electricity had been harnessed for continuous motion. Here was the forerunner of the electric motor. From Riebau's back room to a real laboratory—from once-puzzled observation to rigorous, step-by-step analysis, Faraday thrilled to the knowledge that he had made a significant discovery. A discovery that would serve as a benchmark against which to measure competing electromagnetic theories. A discovery that would validate him in the eyes of the scientific establishment.

George Barnard later recalled that he and Faraday danced around the table, circling the miraculous rotator. Then they closed up the lab and headed off to the circus to celebrate. "I shall never forget the enthusiasm expressed in his face and the sparkling in his

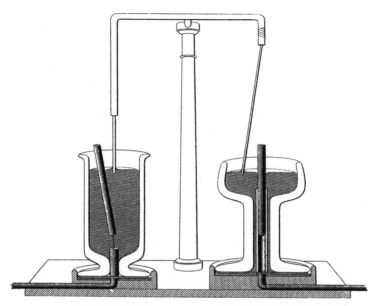

Cross-section of a six-inch-tall demonstration version of Faraday's electromagnetic rotator, precursor of the modern electric motor. When electricity passes through the cups of mercury, the tethered cylindrical magnet revolves around the fixed metal rod (left) and the suspended metal rod revolves around the fixed cylindrical magnet (right).

eyes," George wrote. None of Faraday's glee—or hope for the future—made its way into his laboratory diary, where he soberly summed up the day's work: "Very satisfactory, but make a more sensible apparatus."

For Faraday, the coming months would see the publication of his review article, "A Historical Sketch of Electromagnetism"; the distribution of palm-sized, glass-encased rotators to influential European scientists; the assembly of a large rotator for use in the lecture hall; refinement of the device so that it swung purely in response to the earth's subtle magnetic field; and success where Wollaston had failed—getting a wire to spin around its long axis in response to a

Palm-size version of Faraday's electromagnetic rotator, partly filled with conducting mercury (shaded). When electricity is applied to the pins at the ends of the glass tube, the suspended metal rod revolves around the cylindrical magnet.

magnet. (Ampère had already done it in France.) But, for now, one goal was paramount: to rush his results into print, for priority of discovery in science goes to the claimant who publishes first.

"On some new Electro-Magnetical Motions, and on the Theory of Magnetism" appeared in the *Quarterly Journal of Science* on October 1, 1821. Barely a week later and still basking in the glow of discovery, Faraday realized that he had made a devastating misstep. In his haste to publish, he had neglected to mention Wollaston's preliminary rotation study. Nor had he given politically expedient thanks to his direct superior, Humphry Davy. Instead of being heralded for his achievement, Faraday stood accused of the most unforgivable of intellectual sins: plagiarism.

"I hear every day," Faraday wrote on October 8 to his friend, James Stodart, "more and more of those sounds which though only whispers to me are I suspect spoken aloud amongst scientific men and which as they in part affect my honour and honesty I am anxious to do away with or at least to prove erroneous in those parts which are dishonourable to me . . . My love for scientific reputation is not yet so high as to induce me to obtain it at the expence of honor . . ." Faraday tells Stodart of his wish to meet with Wollaston that he might apologize, convince him that his ideas were his own, and coax him to speak out on his behalf. "I am but a young man & without name and it probably does not matter much to science what becomes of me but if by any circumstances I am subjected to unjust suspicions it becomes no one more than him [Wollaston] who may be said to preside over the equity of science to assist in liberating me from them."

By October's end, Faraday screwed up the courage to write to Wollaston: "I am bold enough sir to beg the favour of a few minutes conversation with you on this subject simply for these reasons that I can clear myself satisfactorily—that I owe obligations to you—that I respect you—that I am anxious to escape from unfounded impressions against me—and if I have done any wrong that I may apologise for it." Wollaston replied immediately: "You seem to me to labour under some misapprehension of the strength of my feelings upon the subject to which you allude. As to the opinions which others have of your conduct, that is your concern, not mine; and if you fully acquit yourself of making any incorrect use of the suggestions of others, it seems to me that you have no occasion to concern yourself much about the matter." Wollaston visited Faraday at the Royal Institution laboratory in early 1822, when Faraday offered to make amends by crediting Wollaston's early work in his next paper. Wollaston evidently declined for his own sake, although he claimed to have advised Faraday nonetheless to credit him, if only to mitigate the plagiarism rumors. Faraday recalled no such suggestion and published his paper without mentioning Wollaston.

Faraday was surely comforted that Wollaston was not miffed about the absence of his name in the rotations papers. But instead of publicly defending Faraday, Wollaston was content to watch over the railing while Faraday thrashed about with the sharks. The sharks turned out to be just two in number, but formidable nonetheless: Henry Warburton, Fellow of the Royal Society and a close friend of Wollaston's; and, to Faraday's amazement—or perhaps not—Humphry Davy.

Warburton was incensed at Faraday for having violated the unwritten protocol of crediting one's scientific elders, no matter how peripheral their assistance. He started a whisper campaign among his Royal Society colleagues to express his indignation. He may have found a receptive audience in a cohort of old-guard members, if not in the Society's more liberal wing, who felt that the hidebound organization was too narrowly focused on agricultural and industrial applications, to the detriment of basic research like Faraday's. In the end, Warburton's efforts came to naught. It appeared that Michael Faraday's "sin" would be forgiven, if only for lack of interest. But Humphry Davy refused to let the accusation die. Like Warburton, he felt that Faraday had intruded on territory already staked out by Wollaston. And, now, that error had been compounded by the appearance of yet another paper sans acknowledgment.

In March 1823, Davy delivered a speech to the Royal Society on the subject of electromagnetism, in which he reminded the gathered members that he had witnessed Wollaston's early experiments in electromagnetic rotation. A reporter's account of Davy's talk, published in the *Annals of Philosophy,* included these lines: "Had not an experiment on the subject made by Dr. W[ollaston] in the laboratory of the Royal Institution, and witnessed by Sir Humphry, failed *merely through an accident* which happened to the apparatus, *he would have been the discoverer of the phenomenon.*" When Faraday objected to Davy's insinuation that Faraday was claiming undeserved credit for the discovery of electromagnetic rotation, Davy disavowed the report, saying he had been misquoted. (The reporter assured Faraday that his account accurately reflected what Davy had said.) An erratum appeared in the journal, but the damage had been done. Davy had rekindled the plagiarism issue.

To Ampère, Faraday privately complained, "I am compelled to say I have not found that kindness, candour and liberality at home which I have now on several occasions uniformly experienced from the Parisian men of Science ... Considering the very subordinate position I hold here and the little encouragement which circumstances hold out to me I have been more than once tempted to resign scientific pursuits altogether ... I struggle on in hopes of getting results at one time or another that shall by their novelty or interest raise me into a more liberal and active sphere."

In early 1823, Faraday scored another triumph: the liquefaction of chlorine gas. (Upon learning that Thomas Northmore had already liquefied chlorine in 1805, Faraday withdrew his claim of priority.) Faraday dutifully credited Humphry Davy for suggesting that he try heating a chlorine compound under pressure, a procedure that produced a few precious drops of bright yellow liquid chlorine. But Davy appended a note to Faraday's published paper, as was his custom, claiming to have secretly predicted the experiment's result. Faraday read the note as Davy's attempt to elevate his own role in the discovery—and discount the many treacherous hours Faraday had toiled in the lab. (Explosions were common in the high-pressure apparatus, and more than once Sarah Faraday plucked glass shards out of her husband's skin.) For two years, Faraday had doggedly pursued his personal investigation of chlorine and its compounds. It was *his* project, he felt, sandwiched into what little spare time he had. He resented that Davy could breeze into the lab between engagements, drop a suggestion, and then, by implication, claim Faraday's independent work as his own. The issue would occasionally resurface, as in 1827, when a guest lecturer at the Royal Institution credited Davy with the liquefaction of chlorine. Faraday testily corrected the lecturer, pointing out that *he,* Faraday, had done it, "unaided by any knowledge of Sir H. Davy's views."

On May 3, 1823, Richard Phillips informed Faraday that he had nominated him for Fellowship in the Royal Society: "I spoze you noze as I did your bizzness at the R. S. Did it well I thinks ..."

Phillips had recruited more than two dozen Royal Society members to sign Faraday's nomination certificate, including, significantly, William Hyde Wollaston. The nomination came as no surprise to Faraday (who may have suggested it), although it evidently did to the Royal Society's president, Humphry Davy. On a rough copy of a letter he wrote in 1823 regarding his nomination, Faraday penned at the bottom: "Sir H. Davy angry, May 30th."

Davy viewed Faraday's nomination as a sign of disrespect for the office of president and a challenge to his own authority. Davy's predecessor, the autocratic Joseph Banks, had always been consulted before a candidate was officially considered for membership in the Royal Society. This way (at least in theory), nominations of unqualified candidates could be quietly "vetoed" in advance of the formal procedure. As president, Davy had been trying to stitch together two warring factions within the Royal Society. Many veterans of the Banks era wanted the Society to continue promoting areas of study that advanced agriculture or industry. The reformist element lobbied instead in support of efforts in mathematics and the physical sciences, including those of a purely theoretical nature. Weary of the Society's centralized power structure—Banks had served as president for over forty years—the reformers sought a more democratic approach to governance. Faraday's nomination was their shot across Davy's bow: Turn about or face a divisive and potentially embarrassing fight.

Davy faced a harrowing choice: Support the nomination of Faraday—a plagiarizer!—and subject himself to charges from the conservatives of nepotism and pandering to the reformers; or oppose the nomination and harden the reformers against him. In the end, political expedience—and simmering jealousy over Faraday's rising reputation—trumped any urge to help the acolyte whose career he had so patiently nurtured. In Davy's view, Faraday had made this mess; it was up to Faraday to clean it up.

"Sir H. Davy told me I must take down my certificate," Faraday later recalled. "I replied that I had not put it up; that I could not take it down, as it was put up by my proposers. He then said I must get my proposers to take it down. I answered that I knew they would not do so. Then he said, I as President will take it down. I replied that I was sure Sir. H. Davy would do what he thought was for the good of the Royal Society."

With Davy resolute in his opposition, Faraday tried instead to sway his other detractor, Wollaston's friend, Henry Warburton. On May 30, 1823, Faraday urged Warburton to hear him out in person. "You would probably find yourself engaged in doing justice to one who cannot help but feel that he has been injured though he trusts unintentionally. I trust that you are not in possession of all the circumstances of the case but I am also sure you will not wish willingly to remain ignorant of them." The two men met on June 5. Faraday recited the facts as he saw them: that he had stopped at Wollaston's house before submitting his paper on electromagnetic rotation, but Wollaston had been away; that he had deemed it presumptuous to describe Wollaston's unpublished work without permission; that the discussions he had overheard between Wollaston and Davy had in no way stimulated his own studies.

A month later, Faraday's detailed time line of his discovery of electromagnetic rotation appeared in the *Quarterly Journal of Science*. Here, too, he acknowledged that he was "M.," author of the highly regarded electromagnetism review articles in the *Annals of Philosophy;* clearly, his mind had ranged far beyond the scope of Wollaston's crude experiment. Upon reading the new paper, Warburton told Faraday that he had never intended to derail his acceptance to the Royal Society; it was Faraday's conduct he objected to, and he had stated as much in private to colleagues—including Davy. But he was now satisfied with Faraday's explanation of events. "When I meet with any of those in whose presence such conversation may have passed," Warburton assured Faraday, "I shall state that my objections to you as a Fellow are & ought to be withdrawn, & that I now wish to forward your election."

Faraday's nomination was announced at ten successive Royal Society meetings and, at the eleventh, on January 8, 1824, secret ballots were cast. Faraday was elected—with one dissenting vote. Of his troubled path to Fellowship, Faraday remarked in his later years that it had been "sought and paid for."

Neither Faraday nor Davy ever spoke publicly about the Royal Society debacle nor, it appears, did they ever have a "heart-to-heart" about its impact on their relationship. Their correspondence ended in the midst of the discomfiting nomination campaign with a brief instruction from Davy about the forwarding of his mail—and a wish for Faraday's "health and success." In 1825, Davy initiated Faraday's promotion to director of the laboratory at the Royal Institution, a more fitting title for a Fellow of the Royal Society. (The Institution's largesse did not extend to salary, which stayed pegged at the original 100 pounds a year until 1853, when it was raised to 300 pounds.)

Chemist John Tyndall, who eventually succeeded Faraday at the Royal Institution, sagely observed, "Brothers in intellect, Davy and Faraday, however, could never have become brothers in feeling; their characters were too unlike. Davy loved the pomp and circumstance of fame; Faraday the inner consciousness that he had fairly won renown. They were both proud men. But with Davy pride projected itself into the outer world; while with Faraday it became a steadying and dignifying inward force."

Davy was surely aware of his own obligation, not merely to instruct, but to foster his protégé's independence. While traveling through the Scottish Highlands in August 1821—the very eve of Faraday's breakthrough discovery of electromagnetic rotation—Davy's poetic muse was stirred by the sight of eagles teaching their young to fly:

The mighty birds still upward rose
In slow but constant and most steady flight,
The young ones following; and they would pause,
As if to teach them how to bear the light,
and keep the solar glory full in sight.
. . .
Their memory left a type and a desire:
So should I wish towards the light to rise,
Instructing younger spirits to aspire
Where I could never reach amidst the skies,
And joy below to see them lifted higher,
Seeking the light of purest glory's prize.

When finally challenged by Faraday, Davy was somehow powerless to translate his poetic ideals into concrete actions. Impelled
by the swift winds of politics, society, and envy, Davy tried to block
his own protégé from "rising to the light." John Tyndall wrote of the
situation, "A father is not always wise enough to see that his son has
ceased to be a boy, and estrangement on this account is not rare;
nor was Davy wise enough to discern that Faraday had passed the
mere assistant stage, and become a discoverer." Davy's brilliant career was already in full ebb when he contracted a debilitating illness
in 1826. He resigned from the presidency of the Royal Society and,
seeking a more therapeutic climate, embarked on a series of extended tours of Europe—mostly without his wife. Davy died in
Geneva in 1829 at the age of fifty-one.

After Davy's death, Faraday sent biographer John Ayrton Paris
the first letter he ever received from Davy, in which Davy offered to
help the dreamy young man in any way he could. With the letter,
Faraday enclosed the instruction, "I send you in the original, re-
questing you to take great care of it, and to let me have it back, for
you may imagine how much I value it." Faraday remained grateful
to Davy for plucking him from obscurity and turning him into a

capable scientist. Whatever bitterness he harbored against his former mentor, he kept private, as was usual for him in such matters. Perhaps the closest he ever came to revealing his sense of betrayal was in a letter to Ampère, who was being vilified by jealous colleagues: "I never yet even in my short time knew a man to do anything eminent or become worthy of distinction without becoming at the same time obnoxious to the cavils and rude encounters of envious men. Little as I have done, I have experienced it and that too where I least expected it."

To Faraday, Davy was surely a consummate example of the fallible man, that "singular compound" of paradoxes and contradictions upon which Faraday had once mused to Benjamin Abbott. Even so, Faraday had right to feel aggrieved at Davy, not only for stifling his drive toward independence, but also for forcing him, against his Sandemanian sensibilities, to press for his own advancement. Humility was one of the defining elements of Faraday's character. But his was a humility tempered by pragmatism. When Davy opposed him, he did what he had to do, and subdued his internal tempests on Sundays at church. Although humble, Faraday was not averse to standing his ground. (Lady Davy comes to mind.) John Tyndall once chided him for signing a letter, "Humbly yours," to which Faraday replied, "Well, but, Tyndall, I *am* humble; and still it would be a great mistake to think that I am not also proud."

The perceptive Tyndall also came to recognize a side of Faraday that those less acquainted with the man might have missed. "We have heard much of Faraday's gentleness and sweetness and tenderness. It is all true, but it is very incomplete. You cannot resolve a powerful nature into these elements, and Faraday's character would have been less admirable than it was had it not embraced forces and tendencies to which silky adjectives 'gentle' and 'tender' would by no means apply. Underneath his sweetness and gentleness was the heat of a volcano. He was a man of excitable and fiery nature, but through high self-discipline he had converted the fire into a central glow and motive power of life, instead of permitting it to waste itself in useless passion. 'He that is slow to anger,' saith the sage, 'is greater than the mighty, and he that ruleth his own spirit than he that taketh a city.' Faraday was *not* slow to anger, but he completely ruled his own spirit, and thus, though he took no cities, he captivated all hearts."

Privately, Faraday must have assessed his own potential as at least equal to, if not greater than, that of the dissipated Davy. And his anger at Davy would therefore seem entirely justified. Still, it could only have been wrenching to be rejected by the man who was inarguably his "father" in science. Of their rift, he would only say, "I was by no means in the same relation to scientific communication with Sir Humphry Davy after I became a Fellow of the Royal Society as before that period; but whenever I have ventured to follow in the path which Sir Humphry has trod, I have done so with respect and with the highest admiration of his talents."

Faraday is said to have remarked that Davy served as "a model to teach him what he should avoid." It's hard to imagine the circumspect Faraday uttering such a statement in a spirit of contempt. On the contrary, viewed in the context of Faraday's life, the remark makes eminent sense. Faraday never mentored a student as Davy had once mentored him; he avoided the high offices that Davy had so eagerly sought; he was content with his midlevel position in English society, a station Davy had desperately tried to escape. Not that Davy's example led Faraday to adopt these facets of his life. That had more to do with Faraday's solitary working style and his obsessive pursuit of research to the exclusion of all but the most essential of life's duties and pleasures. No, Davy's was a cautionary tale—not unlike the biblical parables Faraday loved—about the folly of human arrogance. And yet, flawed as he was, Davy was also the brilliant scientist with whom Faraday had worked shoulder to shoulder, and whose lightning inspiration had flowed to him as surely as if they had been one.

Part of Faraday would remain ever in Davy's debt. But now, having emerged from under the great man's shadow, Faraday could see for the first time, crystal clear, the future he had so long imagined. Nature beckoned him. And, at this moment, it seemed simply a matter of deciding where to begin. Yet, as Michael Faraday was about to learn, freedom is relative. He had merely exchanged one set of shackles for another.

6

HE SMELLS THE TRUTH

Success is the ability to go from one failure to another with no loss of enthusiasm.

—WINSTON CHURCHILL

With his newly acquired Fellowship in the Royal Society, Faraday was now an acknowledged member of the scientific establishment, with all its privileges—and obligations. Humphry Davy, Royal Society president, stood ready to ensure that the fledgling Fellow discharged his responsibilities to government and society. Even as his nomination was being debated, Faraday acceded to Davy's request to help him develop an electrochemical process to prevent corrosion of the copper-sheathed hulls of British warships; the project was a failure. (Barnacles and seaweed adhered to the modified surface.) Davy also invited Faraday to become secretary of the new Athenaeum Club, a social organization for the scientific and literary intelligentsia. The position—with all its list-making, record-keeping, and letter-writing—proved to be a throwback to Faraday's former valet days, and he resigned after a few months. Davy next nominated Faraday to membership on a government-sponsored committee to find ways to improve the production of high-quality optical glass for telescopes. Added to this were Faraday's time-consuming efforts to stem the swelling tide of red ink that threatened to sink the Royal Institution. And his day-to-day management of the house and laboratory, including building maintenance, supervision of servants, and purchase of everything from seat cushions to crucibles. And the seemingly endless stream of visitors seeking technical advice. Plus his ongoing assistance at Brande's lectures.

By 1825, Faraday was up to his neck in a morass of competing responsibilities. He bemoaned his situation to his friend Ampère: "Every letter you write me states how busily you are engaged and I cannot wish it otherwise knowing how well your time is spent. Much of mine is unfortunately occupied in very common place employment and this I may offer as an excuse (for want of a better) for the little I do in original research." Nevertheless, Faraday's laboratory diaries reveal that he managed to squeeze in a surprising amount of experimentation during this time, most notably the discovery of benzene, a liquid compound of carbon and hydrogen, which later became an essential ingredient in organic chemistry and the dye industry. (The new substance was a byproduct of whale oil and codfish oil distillation, whose gas was bottled and sold for illumination. Faraday's older brother Robert worked for the Portable Gas Company in London.)

Faraday's first order of business, now as director of the laboratory, was to rescue the ailing Royal Institution—his home and workplace—from its perpetual money woes. Although burdened by the Institution's demands, he recognized that his own well-being and his effectiveness as a researcher hinged on the viability of the Institution. Already Brande had taken a significant pay cut, and there was always grief about the winter coal bill. (The 1816 bill was settled two years late.) In 1823, the Institution might have had to close its doors had it not been for a no-interest loan from members. No doubt, income from Faraday's lucrative commercial analyses helped keep the Institution afloat for a while; but every minute spent in service to a business customer was stolen from his time for original research. What the Royal Institution needed—and what *he* needed—was a steady source of funds that did not simultaneously hobble his long-term aspirations. In nineteenth-century England, other than the rare generous donor, there was only one such well-spring of money: the public. If Humphry Davy, in his heyday, could draw hundreds of paying patrons into the lecture hall, then surely, Faraday believed, so could he.

Faraday launched his fund-raising campaign in 1826, periodically inviting Institution members into the laboratory, where he regaled them with news of exciting developments in science. What better way to put members in a giving frame of mind than to have them stand at the altar of the Institution's very reason for being. Word of these casual Friday evening soirées spread. By April 1827, the crowd had grown sufficiently large that Faraday moved his presentations and illustrative experiments to the lecture hall upstairs. All his oratory lessons, speaking experience at the City Philosophical Society, and critical self-reflection now came to the fore in the larger venue. Now he himself stood at the vaunted U-shaped table where Humphry Davy, ubermeister of the motivational lecture, had pried open purses with flaming rhetoric. Faraday knew he was no Davy. He could never spout the literary ruffles and flourishes that his former mentor had once used to titillate his rapt audiences. As in all manner of self-expression, Faraday was, always had been, and always would remain prose to Davy's poetry. But *such* prose—a steady, luminous candescence of ideas that illuminated nature as brilliantly as his predecessor's verbal fireworks. Faraday became his audience's unpretentious guide to the scientific realm by simply conveying his own infectious sense of wonder. Listeners left the hall, if not with true understanding, at least with a deeper appreciation of what scientists do—and why they should be supported.

With the move upstairs was born the Royal Institution's Friday Evening Discourses, which Faraday oversaw until 1862 (and which continue to this day). Faraday's purpose in conducting the Discourses is telegraphed in some early lecture notes: "Evening opportunities—interesting, amusing; instruct also;—scientific research—abstract reasoning, but in a popular way—dignity;—facilitate our object of attracting the world, and making ourselves with science attractive to it." Thus Faraday had a broader goal than just institutional fundraising: "I am persuaded that all persons may find in natural things an admirable school for self-instruction, and a field for the necessary mental exercise; that they may easily apply their habits of thought, thus formed, to a social use; and that they ought to do this, as a duty to themselves and their generation."

Faraday found an ally in William Jerden, editor of London's popular *Literary Gazette,* which had featured Faraday's work as early as 1818. The *Gazette* unabashedly highlighted the nation's artistic, literary, and scientific pursuits; indeed, Jerden's working principle was "praise heartily . . . censure mildly." Now Jerden described Faraday as "one of the most successful chemical enquirers of the age" and did his utmost to promote the Royal Institution and its Friday Discourses: "This meritorious and valuable institution has fortunately been raising itself, during the last two or three years, from a state of such depression, into which untoward circumstances, and some little want of energy, has conspired to throw it. The lectures now in the course of delivery, are, and deserve to be numerously attended; and the Friday Evening Meetings are at once the most rational and pleasurable assemblies which are to be found in London." The Discourses quickly became the Friday evening place to be—and to be seen—among London's moneyed class. Faraday's lectures especially, with their sometimes rollicking visual demonstrations, secured his reputation as England's scientific ambassador to its citizenry—or at least to those of means. And the Royal Institution's coffers began to fill.

The majority of the Discourses were delivered by invited speakers; Faraday himself presented 123 Discourses during his career,

and these involved all the behind-the-scenes preparation and re-
hearsal of a stage production–which indeed they were. Faraday
and his fellow lecturers reliably attracted audiences of 400 or more,
up to the hall's capacity of about a thousand; his own 1851 Discourse on atmospheric magnetism drew 1,028 guests. Lectures
started at 9 P.M. sharp, lasted precisely an hour, and then adjourned
for tea and discussion in the library, where various types of apparatus and specimens of interest were displayed. Among the topics of
discussion: a proposed tunnel under the Thames; reproduction in
planaria worms; lighthouses; electrically powered looms; dry rot;
artificial rubies; magnetic force; mine ventilation; reflective coatings
on glass; manufacture of pens; and the mysterious vibration of mi-
croscopic particles known as Brownian motion. The roster of invited lecturers includes John Dalton, on the theory of atoms;
Charles Lyell, on the age of volcanoes; Thomas Huxley, on the origin of species and races; William Thompson, Lord Kelvin, on the
motive power of heat engines; Richard Owen, on the gorilla; Hermann von Helmholtz, on the conservation of forces; and James
Clerk Maxwell, on the theory of primary colors.

One of Faraday's avowed goals for the Discourses was to instruct the public, not just about the results of science, but about
the process itself. "The *laws of nature,* as we understand them, are
the foundation of our knowledge in natural things. So much as
we know of them has been developed by the successive energies of
the highest intellects, exerted through many ages. After a most rigid
and scrutinizing examination upon principle and trial, a definite expression has been given to them; they have become, as it were, our
belief or trust. From day to day we still examine and test our expressions of them. We have no interest in their retention if erroneous; on the contrary, the greatest discovery a man could make
would be to prove that one of these accepted laws was erroneous,
and his greatest honour would be the discovery. Neither should
there be any desire to retain the former expression:–for we know

that the new or amended law would be far more productive in results, would greatly increase our intellectual acquisitions, and
would prove an abundant source of fresh delight to the mind."

One regular attendee on Friday evenings was Jane Marcet, author of *Conversations in Chemistry,* whom Faraday referred to as his
"first instructress." Friedrich von Raumer, a German polyglot traveling in England, commented on Faraday's delivery: "He speaks
with an ease and freedom . . . as some learned professors do; he delivers himself with clearness, precision and ability. Moreover, he
speaks his language in a manner which confirmed me in a secret
suspicion I had, that a number of Englishmen speak it very badly.
Why is it that French in the mouth of [actress] Mlle Mars, German
in that of [dramatist Johann] Tieck, English in that of Faraday seems
a totally different language?–because they articulate what other
people swallow and chew." Among London's glitterati, Lady Holland burst forth with effusive, if hyperbolic, praise: "It was an irresistible eloquence, which compelled attention and insisted upon
sympathy. It waked the young from their visions and the old from
their dreams . . . His enthusiasm sometimes carried him to the point
of ecstasy when he expatiated on the beauty of nature . . . A pleasant vein of humour accompanied his ardent imagination, and occasionally, not too often, relieved the tension of thought imposed
upon his pupils. He would play with his subject now and then, but
very delicately; his sport was only just enough to enliven the effort
of attention." And chemist William Crookes spoke for the professionals in the audience when he remarked, "All is a sparkling stream
of eloquence and experimental illustration. We defy a chemist who
loves his science, no matter how often he may himself have repeated
an experiment, to feel uninterested when seeing it done by Faraday."

John Tyndall noted that people left the hall satisfied, even if they
did not grasp the nuances of Faraday's lecture: "[H]is Friday evening discourses were sometimes difficult to follow. But he exercised

a magic on his hearers which often sent them away persuaded that they knew all about a subject of which they knew but little." (Modern-day teachers lament this same effect in their students all the time.) Cornelia Crosse, wife of science enthusiast Andrew Crosse, remarked, "No attentive listener ever came away from one of Faraday's lectures without having the limits of his spiritual vision enlarged, or without feeling that his imagination had been stimulated to something beyond the mere exposition of physical facts." The Friday-night Faraday was scientist as showman, shaman, and storyteller. And the great U-shaped table was the "hearth" about which the tribe gathered to hear tales of the natural world.

It had been fourteen years since Faraday had conveyed to his friend, Benjamin Abbott, his initial observations on the art of lecturing, and he periodically added to his collection of ideas. Faraday's advice to the novice—and experienced—lecturer is as relevant today as it was in his own time:

A flame should be lighted at the commencement [of the lecture] and kept alive with unremitting splendour to the end.

If at a loss for a word, not to ch-ch-ch- or eh-eh-eh, but to stop and wait for it. It soon comes . . .

A lecturer should appear easy & collected, undaunted & unconcerned, his thoughts about him and his mind clear and free for the contemplation and description of his subject . . . His whole behaviour should evince respect for his audience and he should in no case forget that he is in their presence.

Apt experiments . . . ought to be explained by satisfactory theory, or otherwise we merely patch an old coat with new cloth, and the whole (hole) becomes worse.

A lecturer falls deeply beneath the dignity of his character when he descends so low as to angle for claps & asks for commendation . . .

'Tis well too when a Lecturer has the ready wit and the presence of mind to turn any casual circumstance to an illustration of his subject. Any particular circumstance that has become table talk for the town, any local advantages or disadvantages, any trivial circumstance that may arise in company give great force to illustrations aptly drawn from them and please the audience highly as they conceive they perfectly understand them.

Faraday's critique extended to the physical aspects of the lecture hall: seating arrangements, entry and exit, lighting and, especially, ventilation. "[H]ow often have I felt oppression in the highest degree when surrounded by a number of other persons and confined in one portion of air. [H]ow have I wished the Lecture finished, the lights extinguished, and myself away merely to obtain a fresh supply of that element." Among the Faraday-inspired improvements to the Institution's lecture hall was the conversion of odorous oil lamps to cleaner-burning gas.

Faraday extended the Royal Institution's outreach effort by initiating an annual Christmas lecture series for children. The first "juvenile" lectures, on astronomy, were given in 1826 by Faraday's friend, Professor J. Wallis. In 1827, Faraday himself delivered a six-lecture series on chemistry, and in subsequent years, eighteen others on topics such as electricity, combustion, and common metals. Faraday made sure his young listeners were entertained while they learned: He wrote on paper using an electric pen; burst iron bottles with freezing water; burned gunpowder in water and pieces of iron in alcohol; exploded a hydrogen-filled balloon with an electric spark; created an arch of iron filings above a concealed horseshoe magnet; made his hair stand on end from accumulated electric

Faraday delivering one of his Christmas Juvenile lectures at the Royal Institution.

charge; and ignited a gas jet with a spark from his finger. Part of Faraday's effectiveness before young audiences was his obvious delight in what he was doing. He was everyone's kindly, indulgent uncle, with a sparkle of youthful wonder that never deserted him. In his later years, the annual Christmas lectures were to become a source of renewal: "I will return to second childhood, and become, as it were, young again among the young."

Faraday's lectures were laced with adjectives like *wonderful* and *beautiful.* "Look at these colours," he told an audience while projecting light refracted in a crystal. "[A]re they not most beautiful for you and for me? (for I enjoy these things as much as you do.)" He tried to awaken the rational inquirer in every child, in the hope that some of them might grow up to be scientists: "Always remember that when a result happens, especially if it be new, you should say,

'What is the cause? Why does it occur?' and you will in the course of time find the reason." And to the children, he distilled the experimental method to its very essence: "[I]n the pursuit of science we first start with hopes and expectations; these we realize and establish never again to be lost, and upon them we found new expectations of further discoveries, and go on pursuing, realizing, establishing, and founding new hopes again and again."

Charles Dickens was sufficiently impressed by Faraday's approach that he twice urged Faraday to expand his lecture notes into publishable form. Faraday declined, claiming that a written version would lack the immediacy and educational force of the direct experience. However, transcriptions of two of his lectures, "The Chemical History of the Candle" (1848 and again in 1860) and "Lectures on the Various Forces of Matter" (1859), were completed with Faraday's consent.

Faraday begins "The Chemical History of the Candle" by introducing various types of candles that were displayed at the front of the hall. He next describes the structure, variation, and brightness of the flame and the action of the wick. He discusses over several lectures the various gases involved in combustion (air, hydrogen, oxygen, carbon dioxide), plus byproducts such as water and carbon. He uses familiar toys to illustrate concepts of gas density and pressure—hot-air balloon, popgun, blowpipe, suction cup—saying, by way of explanation, "We young ones have a perfect right to take toys, and make them into philosophy, inasmuch as now-a-days we are turning philosophy into toys." He points out that respiration is quite literally a living process of combustion analogous to that of a burning candle. (Oxygen we breathe reacts with digested food—"fuel"—within the bloodstream, generating carbon dioxide which we expel.) Faraday concludes on an uplifting note: "All I can say to you at the end of these lectures (for we must come to an end at one time or another) is to express a wish that you may, in your generation, be fit to compare to a candle; that you may, like it, shine as light to those about you; that in all your actions, you may justify the beauty of the taper by making your deeds honorable and effective in the discharge of your duty to your fellow-men."

In "Forces of Matter," Faraday invites the children to share his enduring sense of wonder at the seemingly ordinary: "Let us now consider, for a little while, how wonderfully we stand upon this world. Here it is we are born, bred, and live, and yet we view these things with an almost entire absence of wonder to ourselves respecting the way in which all this happens . . . [W]ere it not for the exertions of some few inquiring minds, who have looked into these things and ascertained the very beautiful laws and conditions by which we do live and stand upon this earth, we should hardly be aware that there was anything wonderful in it." He proceeds through a host of seemingly dissimilar forces—simple pushes and pulls, electrical attraction and repulsion, gravity, cohesion, chemical affinity—and only then reveals their interconnectedness and underlying unity, all to the accompaniment of dozens of demonstrations. One attendee recalled: "He made us all laugh heartily; and when he threw a coalscuttle full of coals, a poker, and a pair of tongs at the great magnet, and they stuck there, the theatre echoed with shouts of laughter." Faraday closes with his trademark appeal to budding scientists, this time with an assist by Shakespeare:

[W]hat study is there more fitted to the mind of man than that of the physical sciences? And what is there more capable of giving him an insight into the actions of those laws, a knowledge of which gives interest to the most trifling phenomenon of nature and makes the observing student find

Tongues in trees, books in running brooks,
Sermons in stones, and good in every thing?

As a boy, English shipbuilder and philanthropist Alfred Yarrow was one of those inspired by Faraday's Christmas lectures. "At the end of the lectures at the Royal Institution," Yarrow recalled in 1928, "Faraday used to stop sometimes for quite half an hour talking to a lot of us boys, and sometimes making us go through some of his own experiments with our own hands. A friend of mine of the name of Walter Rutt and I were so much impressed with Faraday's personality, that it was our practice on Sundays to meet Faraday and his wife at the corner where he used to go from the Caledonian Road up the hill to the road leading to Barnsbury, where at the corner we would stop and wait to raise our hats to him, which salute he always graciously returned. We then ran down a side street and then one parallel to the one taken by Faraday and his wife and up another street which intersected the road which Faraday passed over, and we saluted him a second time. We then frequently followed him to his chapel . . . This was our regular job on Sunday afternoons."

Workaholic that he was, Faraday nevertheless enjoyed respites from the business of science. The Royal Institution was where he both lived and worked, and these disparate aspects of his life inevitably mingled within the Institution's walls. The upstairs apartment that was his refuge from daily toil was frequently given over to professional reading, writing, and entertaining. Likewise, the downstairs was never exclusively reserved for experimentation or lectures. On occasion, Faraday might be seen barreling down the corridor on a velocipede. The famed laboratory served also as the family manufactory of soda, ginger wine, and lavender lozenges. Relatives visited often, and several nieces stayed with their Uncle Mike and Aunt Sarah for extended periods. Their reminiscences paint a picture of a man equally attentive to his family as to his work.

"After I went, in 1826, to stay at the Royal Institution," recalled Margery Reid, daughter of Sarah's eldest sister, "when my aunt was going out (as I was too little to be left alone), she would occasionally take me down to the laboratory, and leave me under my uncle's eye, whilst he was busy preparing his lectures. I had of course to sit as still as a mouse, with my needlework; but he would often stop and give me a kind word or a nod, or sometimes throw a bit of potassium into water to amuse me [with its fiery ignition]. In all my childish troubles, he was my never-failing comforter, and seldom too busy, if I stole into his room, to spare me a few minutes; and perhaps when I was naughty and rebellious, how gently and kindly he would win me round, telling me what he used to feel himself when he was young, advising me to submit to the reproof I was fighting against."

Faraday enjoyed escapes to the country. A thirty-mile hike was not uncommon for him, even into old age, of course noting flora, fauna, and landforms along the way. Vacationing with relatives, he played charades and ball games, swam in the ocean, and rowed in an eight-oar "cutter." Niece Margery writes of a month spent at her aunt and uncle's cottage in Walmer, in southeastern England, not far from the coast. "How I rejoiced to be allowed to go there with him! We went on the outside of the coach, in his favourite seat behind the driver. When we reached Shooter's Hill, he was full of fun about Falstaff and the men in buckram, and not a sight or sound of interest escaped his quick eye and ear . . . In those days I was eager to see the sun rise, and my uncle desired me always to call him when I was awake. So, as soon as the glow brightened over Pegwell Bay, I stole downstairs and tapped at his door, and he would rise, and a great treat it was to watch the glorious sight with him . . . As he had gone to Walmer for rest and refreshment, I, the young one of the party, had to inveigle him away from his books whenever I could. Sometimes I was allowed to go to read with him, and my grandmother, who was staying with us, used to say, 'What sort of reading lessons are those going on upstairs? I hear "ha! ha!" more than any other sound.' . . . When anything touched his feelings as he read—and it happened not unfrequently—he would show it not only in his voice, but by tears in his eyes also."

Faraday took every opportunity to teach his nieces, as he once had his younger sister. All subjects were fair game—nature, arithmetic, elocution, and once, Margery recalled, an impromptu lesson in vocabulary: "Nothing vexed him more than any kind of subterfuge or prevarication, or glossing over things. Once I told him of a professor, previously of high repute, who had been found abstracting some manuscript from a library. He instantly said, 'What do you mean by abstracting? You should say stealing; use the right word, my dear.' "

Faraday's devotion to family was absolute, whether in celebration or sorrow. "In times of grief or distress," writes Margery, "his sympathy was always quick, and no scientific occupation ever prevented him from sharing personally in all our sorrows and comforting us in every way in his power. Time, thoughts, purse, everything was freely given to those who had need of them." Faraday extended his critical faculties equally upon the natural world and the world of man, drawing connections between them. It seemed that nature was replete with lessons to illuminate human conduct and bring solace in times of adversity. And, while biblical doctrine was barred from the laboratory, his advice on personal matters shows an easy mingling of reasoned, "scientific" judgment with absolute faith in the benevolence of God. In 1826, while summering in Niton, Faraday composed a response to a melancholy letter from his brother-in-law, Edward Barnard. In consoling Edward, Faraday gently spells out his own philosophy about the vicissitudes of life:

I have been watching the clouds on these hills for many evenings back: they gather when I do not expect them; they dissolve when, to the best of my judgment, they ought to remain; they throw down rain to my mere inconvenience, but doing good to all around; and they break up and present me with delightful and refreshing views when I expect only a dull walk . . . So it is in life; and though I pretend not to have been much involved in the fogs, mists, and clouds of misfortune, yet I have seen enough to know that many things usually designated as troubles are merely so from our own particular view of them, or else ultimately resolve themselves into blessings. Do not imagine that I cannot feel for the distresses of others . . . I do feel for those who are oppressed either by real or imaginary evils, and I know the one to be as heavy as the other. But I think I derive a certain degree of steadiness or placidity amongst such feelings by a point of mental conviction, for which I take no credit as a piece of knowledge or philosophy, and which has often been blamed as mere apathy. Whether apathy or not, it leaves the mind ready and willing to do all that can be useful, whilst it relieves it a little from the distress dependent upon viewing things in their worst state. The point is this: in all kinds of knowledge I perceive that my views are insufficient, and my judgement imperfect. In experiments I come to conclusions which, if partly right, are sure to be in part wrong; if I correct by other experiments, I advance a step, my old error is in part diminished, but it is always left with a tinge of humanity, evidenced by its imperfection. The same happens in judging of the motives of others; though in favorable cases I may see a good deal, I never see the whole. In affairs of life 'tis the same thing: my views of a thing at a distance and close at hand never correspond, and the way out of a trouble which I desire is never which really opens before me. Now, when in all these, and in all kinds of knowledge and experience, the course is still the same, ever imperfect to us, but terminating in good, and when all events are evidently at the disposal of a Power which is conferring benefits continually upon us, which, though given by means and in ways we do not comprehend, may always well claim our acknowledgment at last, may we not be induced to suspend our dull spirits and thoughts when things look cloudy, and, providing as well as we can against the shower, actually *cheer our spirits* by the thoughts of the good things it will bring with it? and will not the experience of our past lives convince us that in doing this we are far more likely to be right than to be wrong?

Faraday's "steadiness" and "placidity" were most sorely tested by his government-backed research on optical glass. Throughout much of the 1700s, England had held a virtual monopoly on the manufacture of the specialized glass required for lenses in telescopes, theodolites, and other optical instruments. However, a confiscatory excise tax so decimated the English glass industry that by the late 1820s the government was forced to launch a crash program to rediscover the means of production. The import market for the prized flint glass was nonexistent; other countries used virtually every ounce produced within their borders. By the 1820s, German opticians were making superb nine-inch-diameter lenses, at least twice as large as the best English lenses.

Faraday had felt duty-bound in 1825 to accept Davy's invitation to join the Committee for the Improvement of Glass for Optical Purposes—in fact, to effectively manage the project—not only because of his patriotism but also because of the revenue it would bring to the Royal Institution. Experiments were initially conducted at the Falcon Glass Works, three miles from the Royal Institution, until 1827, when Faraday demanded that a glass furnace be installed in his laboratory. The work became a quagmire of tedium, repeated failure, and exhaustion. If the furnace was too cool, air bubbles infiltrated the finished glass. If the furnace was too hot, heavier ingredients sank, rendering the glass non-uniform. If the stirring rods were made improperly, contaminants leached into the molten glass. Chemical reactions occurred that clouded the glass when it set. On occasion, the bottoms of the melting crucibles liquefied and fused to the floor of the furnace. Each series of trials provoked another series, with only incremental progress.

The casual observer of Faraday's efforts in optical glass might be reminded of the former mantelpiece experimenter in Riebau's shop: Try this, try that, try another thing. But repeated, focused trials, each modified ever so slightly from the previous one, became the hallmark of Faraday's exhaustive (and sometimes exhausting) approach to laboratory investigation. Faraday tried to examine every conceivable facet of a phenomenon in a patently obsessive attempt to uncover everything of consequence. In searching for the roots of an effect, he hoed the entire field, not just where stems poked up from below. Royal Institution colleague John Tyndall remarked perceptively that Faraday's experimental method "united vast strength with perfect flexibility. His momentum was that of a river, which combines weight and directness with the ability to yield to the flexures of its bed. The intentness of his vision in any direction did not apparently diminish his power of perception in other directions; and when he attacked a subject, expecting results, he had the faculty of keeping his mind alert, so that results different from those which he expected should not escape him through preoccupation." German chemist F. W. Kohlrausch put it more succinctly: "Er riecht die Wahrheit." *He smells the truth.* (Evidently, Faraday also "smelled" falsehood, even that brought about by his own hand. In 1832, he collected all his published papers and bound them, along with the preface: "Papers of mine published . . . since the time that Sir H. Davy encouraged me to write the 'Analysis of Caustic Lime.' Some I think (at this date) are good, others moderate, and some bad. But I have put *all* into the volume, because of the utility they have been to me, and none more than the bad, in pointing out to me . . . the faults it became me to watch and avoid.")

By the summer of 1828, the only tangible results of Faraday's glass research were his own "nervous headaches and weakness." He had no choice but to ramble the countryside with Sarah for two months until the symptoms subsided. This marked the beginning of an alternating pattern of arduous work and enforced relaxation that Faraday would repeat at intervals throughout the rest of his life.

Finally, in 1830, Faraday produced a small disk of passable optical quality. The Royal Society immediately asked him to scale up the process and produce "a perfect piece of glass of the largest size that his present apparatus will admit, and also to teach some person to manufacture the glass for general sale." Faraday refused, telling Davies Gilbert, president of the Royal Society, "I further wish you most distinctly to understand that I regret I ever allowed myself to be named as one of the Committee. I have had in consequence several years of hard work; all the time that I could spare from necessary duties (and which I wished to devote to original research) [has] been consumed in the experiments and consequently given gratuitously to the public. I should be very glad now to . . . return to the prosecution of my own views . . ." Faraday finally resigned from the project on July 4, 1831, sending the Royal Society six volumes of experimental findings and committee minutes. Even a year later, the project remained a touchy subject; in answer to another request to engage in industrial research—this time, involving the manufacture of iron—Faraday wrote: "I had enough of endeavouring to improve a manufacture when I gave all my spare time for nearly three years in working on glass: one such experiment in a mans life is enough." With the cessation of the glass work, the last of Faraday's shackles had been cast off.

7

A TWITCH OF
THE NEEDLE

*Nothing in the world can take the place of persistence. Talent
will not; nothing is more common than unsuccessful men with
talent. Genius will not; unrewarded genius is almost a
proverb. Education will not; the world is full of educated
derelicts. Persistence and determination alone are omnipotent.*

—CALVIN COOLIDGE

By 1831, Michael Faraday had achieved worldwide fame, having
published more than sixty scientific papers and a handbook on laboratory chemistry. He had received honorary appointments to scientific societies throughout Europe. The Royal Society had invited
him to deliver its annual Bakerian lecture in 1829. (He spoke about
optical glass.) He had succeeded in stabilizing the Royal Institution's finances by reaching out to the public through lectures. While
his salary remained a rock-bottom 100 pounds a year, his annual income from commercial chemistry analyses had swelled to 1,000
pounds in 1830, with the prospect of even more in future years.
The minutes of the Royal Institution Managers' Meetings reflect his
rising stature: Initially he is plain "Michael Faraday," then "Mr. Professor Faraday," and finally "Professor Faraday."

Faraday had pondered, and rejected, the offer in 1827 of a chemistry professorship at the new University of London, saying, "The
[Royal] Institution has been a source of knowledge and pleasure to

me for the past fourteen years, and though it does not pay me in Salary for what I *now* strive to do for it, yet I possess the kind feelings & good will of its Authorities & members, all the privileges it can grant (or I require) and moreover, I remember the protection it has afforded me during the past years of my scientific life." Two years later, with the Institution on firmer footing, Faraday did accept a part-time teaching appointment at the Royal Military Academy at Woolrich, but only after receiving assurances that his involvement would not conflict with his ongoing work at the laboratory. "[M]y time is my only estate," he explained to the head of the Academy, "and that which would be occupied in the duty of the situation must be taken from what otherwise would be given to professional business."

During the summer of 1831, freed from the burdens of Davy and optical glass and a shaky Royal Institution, Faraday had to make a crucial decision: Whither his scientific career? Toward the certain riches of a lucrative consulting business? Or toward the equally certain impoverishment of a life devoted to basic research? The choice for him was obvious. He had not come this far only to retreat from his long-held dream. He was an explorer by nature, in blood and brain. Nothing could ever change that. But for the past six years, he had stood on the shoreline while others had sailed into nature's unknown and returned with tales of discovery. Now his own ship was ready to sail. It was only a matter of deciding the direction.

There was a project Faraday had in mind, an idea that had originated a decade earlier when he cobbled together his little electromagnetic rotator. So modest a device, yet simultaneously profound, for its whirring movement revealed the invisible, encircling magnetism generated by an electric current. Magnetism arising from electricity. Would it not be reasonable to assume that the opposite process occurs as well?

On August 29, 1831, Faraday started a new section in his laboratory diary: "Expts. on the production of Electricity from Magnetism, etc. etc."

In the early 1800s, there were two ways to generate electricity: through frictional means, by rubbing together, either by hand or mechanically, a pair of insulating materials to build up static charge; or chemically, via metal-acid batteries. The frictional process was hard to moderate: Accumulated electricity discharged abruptly and uncontrollably. The "voltaic" route did provide steady current for a limited period, but batteries were expensive, noxious, and hard to maintain.

There was a third way to generate a steady electric current—at least in theory. Induction: to draw upon an object's innate electrical or magnetic essence to stimulate current in an adjacent wire. That such a process should work seemed quite logical. After all, a permanent magnet induces magnetism in an iron nail when the two are close together, and a sewing needle becomes magnetized when stroked by a magnet. Similarly, an electrically charged metal globe induces electrical behavior in nearby globes. Might not an electrical current in one wire induce current in an adjacent wire? Or perhaps a magnet might be used in some way to elicit current.

Faraday had already made a few tentative forays into electrical induction. In 1824, he laid a strong magnet inside a current-carrying wire coil, hoping to see some effect on the flow of current. There was none. The following year he placed two circuits side by side so that the live wire might induce current in its passive partner. When that failed, he draped a circuit over an electrically energized coil; here, too, no electricity was induced. Then in 1828, he conducted a series of experiments with a bar magnet and suspended rings of copper, platinum, and silver. Again, no result.

Intermittently, Faraday was seen contemplating a tiny, wire-wrapped iron cylinder he carried in his waistcoat pocket. He knew

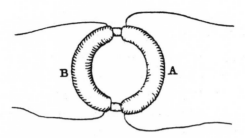

Drawing of Faraday's induction ring.

that sending an electric current into the coiled wire magnetizes the iron—creating an electromagnet. Electricity generates magnetism. But apparently the reverse did not occur. If an ordinary magnet stood in place of the iron cylinder, its magnetism did not stimulate electricity in the surrounding coil. Many had tried to induce electricity from various configurations of magnets or electrified wires. All had failed. And nobody knew why. As the 1820s drew to a close, most scientists were apt to conclude that the induction of electrical current was simply impossible.

A related mystery—dubbed Arago's wheel—had arisen in 1824. French scientist François Arago observed that a disk of *nonmagnetic* metal assumes magnetic properties when it is spun: A magnetized needle suspended from above hangs motionless when the disk is stationary, yet turns when the disk rotates. No one could explain how mere movement could render ordinary metal magnetic. To Faraday, the lingering mysteries meant only one thing: The theoretical foundations of electricity and magnetism were suspect. As he had long maintained, there was no experimental proof of either the fluid model of electricity or Ampère's molecular-current model of magnetism. The time was ripe for a new approach.

On August 29, 1831, with electrical and magnetic ideas having percolated in his mind for a decade, Faraday launched an all-out pursuit of electromagnetic induction. His key investigative instrument was a contrivance of homely elegance: a forged iron ring, six inches in diameter and seven-eighths-inch thick, its opposing halves encased in fabric-covered wire, like a partly shrouded mummy. The ring was essentially a doughnut-shaped version of his pocket electromagnet, but with two separate coils of wire. In winding the wires around the ring, Faraday had meticulously interposed lengths of insulating twine, then sandwiched each succeeding wire layer in cloth. Thus, there was no contact of metal upon metal within each of the two coils: Electricity could spiral around, but not conduct across, the turns of wire; nor could it pass between the coils through the iron ring.

Faraday readied the wire-ends of the "primary" coil (to use modern coinage) for connection to a battery, whose current would transform the enclosed iron ring into an electromagnet. The ring would channel its magnetism into the interior of the "secondary" coil on the opposite side. It was here, in the passive secondary coil, that Faraday sought to induce an electric current from the enclosed magnetism. To discern the feeble flow of current, he extended the wire-ends of the secondary coil over a delicately balanced, horizontal, magnetized needle. A rotation of this needle would indicate the presence of electricity in the wire—and the success of the experiment. (Subsequent experiments involved a more refined version of the indicator needle: a galvanometer.)

Faraday closed the battery connection to the primary coil; the indicator needle twitched, then settled back to its original position. For an instant, a current had arisen in the secondary coil, going in the opposite direction to that in the primary. Faraday opened the battery connection; again the needle twitched, this time the other way. The spurt of secondary current was now in the same direction as the primary current. Between closings and openings of the switch, when the battery's current pumped steadily through the primary

Faraday's induction ring as it appears today.

coil, the needle remained unmoved. No wonder experimenters had missed the effect! Induction occurs only when the battery is being connected to or disengaged from the circuit, that is, only when the primary current and its associated magnetism are *changing*. Indeed, Faraday mused, it was very much as though a *wave* of electrical power had surged through the secondary coil, and almost immediately subsided. Had he not been looking at the galvanometer at just the right instants, its subtle twitching might have gone unnoticed. Faraday's modest iron ring marked the genesis of the electrical transformer, an essential element in the modern electric power industry and in many small consumer devices. (Faraday subsequently learned that "self-induction" occurs even between individual loops of a single coil—a phenomenon discovered almost simultaneously by physicist Joseph Henry in the United States. A buildup of current in one loop will induce a *reverse* current in the next loop; thus, a coil takes time to fully energize. The Royal Institution's great electromagnet—522 feet of copper wire wrapped around a link from a ship's anchor chain—takes two seconds to ramp up to maximum strength.)

Nowhere in Faraday's voluminous writings is there any mention of why he resumed his induction experiments at this time. Or why he knew to be attentive to the galvanometer at the proper moments. In mid-August 1831, he had ordered the iron ring to be cast, then had patiently coiled the wires around it himself. Whereas he and many others had repeatedly failed during the previous decade, he had now succeeded on the very first day of his experiment, as though he had anticipated the correct path. Faraday's research in the years prior to his discovery of induction had been devoted not to electricity and magnetism, but to a diverse variety of activities— and to several alternative, even radical-sounding, ideas.

Between 1827 and 1829, Faraday had read a lengthy translation of Augustin Fresnel's treatise on the wave theory of light. (That light might be a wave, not a particle, had first been proposed by the Royal Institution's Thomas Young in 1801; however, opposition from scientists was so strong that Young renounced the idea.) In Fresnel's model, light emanates from glowing sources as waves of luminous power, like ripples on a pond. The transmission of power takes place without any flow of matter, just a series of local oscillations in an otherwise stationary medium. In other words, the wave's form travels through matter without carrying the matter along. Perhaps, Faraday speculated, electricity is like that: waves of energy that course through a wire, jostling molecules momentarily as they pass, but not sweeping them forward. An "axis of power," he called it somewhat obscurely.

Another idea that may have influenced Faraday's induction experiments had come, oddly enough, from the concert hall. Faraday was a music lover and, between 1828 and 1830, had delivered a series of Friday Evening Discourses on the physics of sound and musical instruments. Again, a wave phenomenon, this time in the form of airborne vibrations. To enhance the lectures, Charles Wheatstone introduced Faraday to Chladni figures: symmetric, stationary patterns created when a plate of sand or powder is rapidly vibrated.

(The vibration was accomplished with a violin bow.) Faraday may have been particularly intrigued by the demonstration that a vibrating plate induces a similar pattern on an adjacent plate, the vibratory power having passed like sound waves through the air. Perhaps he recognized that he had witnessed the audio analog of what he sought to do with electricity: to make one current induce another through some invisible linkage across space. In any event, Faraday became so enamored of the acoustical figures that he conducted an intensive investigation of the phenomenon, working this time with water, egg white, and various oils. During the six months just prior to his induction experiment, he was steeped in the transmission of vibrations and waves, suggesting connections between those he could see or hear and the invisible variety that might arise in the electromagnetic realm: "I am inclined to compare the diffusion of magnetic forces from a magnetic pole, to the vibrations upon the surface of disturbed water, or those of air in the phenomena of sound . . ."

Also in advance of Faraday's breakthrough induction experiment had come news from physicist Gerritt Moll of Utrecht describing a curious feature of electromagnets: When the direction of electricity through the coil of an electromagnet is reversed, the north-south polarity of the magnet follows suit almost instantaneously. Faraday had realized that such a rapid response was not feasible in Ampère's widely accepted model of magnetism; molecular currents take time to align themselves and establish an object's overall magnetic polarity. Once again, the message to Faraday was unmistakable: The theoretical underpinnings of electricity and magnetism were ripe for revision, if not revolution. The guiding light to the future lay in well-planned and -executed experimentation. If not now, when?

The terse account of August 29, 1831, in Faraday's laboratory diary gives no sense of the exhilaration he must have felt seeing the telltale wobble of the indicator needle. Nor does the diary reveal how he felt having to abandon the laboratory a few days later for a planned vacation in Hastings with Sarah. But he used his time away from London to wind a series of new coils for follow-up experiments. From his seaside cottage, Faraday wrote to his friend Richard Phillips, the restrained message surely concealing an urge to pen *Eureka!* "I am busy just now again on Electro-Magnetism and think I have got hold of a good thing but can't say; it may be a weed instead of a fish that after all my labour I may at last pull up."

Faraday resumed his investigation on September 24. Now his aim was different: to induce current, not through *electro*magnetism, as he already had, but with an ordinary magnet. (He had not yet convinced himself that the two magnetisms were identical, so the focus of his induction attempts alternated between them.) Deriving electricity from a magnet, Faraday assumed would be the greater challenge. A magnet's halo of power is more diffuse than that of an electromagnetic coil, and any induced current would likely be harder to detect.

First he brought a bar magnet toward a flat spiral of wire, then yanked it away—the galvanometer needle remained fixed. Next he inserted a wire spiral between a pair of magnetic poles and changed the gap between them—no induction. Then, briefly returning to the electromagnetic side of his investigation, he placed a current-carrying coil and a passive coil side by side—again, no response. When iron coils produced no result, he tried copper—still nothing. (In all these cases, the induction effect was merely too subtle to be registered by the galvanometer.) Finally, toward day's end, he tied together the ends of two bar magnets to form a hinged, V-shaped "jaw," within which he secured a small, upright, coil-enshrouded iron cylinder. The coil-ends he connected to the galvanometer. Every time he opened or closed the magnetic jaw, varying the magnetic state of the iron, the needle flicked: "a mere momentary push or pull," Faraday jotted in his diary; nevertheless, "here distinct conversion of Magnetism into Electricity." It was his only success of the day, yet one of great moment: For the first time electricity had been generated from an ordinary magnet.

Resuming his experiments on October 1, Faraday succeeded again with the iron ring—"powerful effect . . . pulling the needle quite round," he exulted in his diary. He next addressed a crucial point that a lesser researcher might have taken for granted: proving that the induced energy in the secondary coil, *presumed* to be ordinary electricity, *is* in fact ordinary electricity. That is, does the induced energy behave in every way like familiar static and voltaic electricity? Faraday put away the galvanometer. Instead, to the wire-ends of the secondary coil, he attached metal or carbon rods—electrodes. He held the electrodes to his tongue while the induction process was carried out, bracing himself for the expected sting of electricity. It never came. Instead, he plunged the electrodes into various acid solutions, hoping to trigger familiar electrochemical reactions. The solutions remained unaffected. Nor did his most sensitive measurements disclose even the slightest electrical heating of the secondary wire. The spurts of induced current were evidently too brief and too weak to reveal themselves in the usual ways. Fortunately, other electrical behaviors were manifest. The halo of magnetism from the pulses of induced current was sufficient to magnetize needles. And Faraday did manage to coax a minuscule spark from a secondary coil by routing the induced energy into the narrow gap between two carbon electrodes. (When he demonstrated this spark phenomenon to an Oxford audience in 1832, a fusty university dean walked out in protest, muttering, "*Indeed* I am sorry for it; it is putting new arms into the hands of the incendiary.")

On October 17, 1831, Faraday decided to try an even clearer demonstration of induction by "pure" magnetism. Into the cavity of a coil of many turns, he thrust a cylindrical magnet, then quickly withdrew it. In response, the galvanometer needle swung heartily to one side, then to the other. He repeated the action several times to assure himself of the galvanometer's echoing jump. When he thrust the magnet *outside* the coil, the galvanometer remained still. Clearly, the needle was responding to pulses of magnetism from the coil's induced current, not to the thrusts of the magnet itself. Here, simply, indisputably, and at long last was the induction of electric current in its purest form: just a magnet, a coil, and motion. This is what scientists had been seeking for more than a decade.

At the end of October 1831, Faraday packed up his coils and his galvanometer and set out for Woolwich, in greater London, where the Royal Society's great horseshoe magnet was stored. The device consisted of 437 individual bar magnets bound together as one. A one hundred–pound force was needed to wrest an iron bar from its grip, and Faraday wanted to bring this extreme magnetism to bear on his induction experiments. He slipped an iron rod into the cavity of a coil, then eased the rod's ends against the poles of the Woolwich magnet, strongly magnetizing the rod. Every time he broke or reestablished the contact, the galvanometer needle spun wildly before settling back to its original state. Even the rod's approach to the great magnet rotated the needle, as did its withdrawal. The basic rule of induction was the same as before: to induce current, magnetism within a coil must *change*. Constant magnetism, even a magnetic force as powerful as the Woolwich magnet's, will not generate current. Again Faraday had succeeded in producing electricity by purely magnetic means.

As devoted as he was to experiment for experiment's sake, Faraday was certainly attuned to the practical potential of his various magnet-based generators. Such devices, if refined, might provide virtually inexhaustible sources of power for the laboratory—and perhaps, someday, for society at large. Still, their herky-jerky, reciprocating action, whether ultimately powered by muscle or ma-chine, generated energy in pulses only, and was therefore ill-suited to practical use. To effectively power a device—or a house or an en-tire city—electricity has to be continuous.

Faraday's next mechanism, inspired by Arago's wheel, elegantly addressed the idea of practical energy production. Faraday had brought with him to Woolwich a twelve-inch-diameter copper disk that was free to rotate on a brass axle. He secured the disk so that its edge fell between the poles of the big magnet. Then he spun it, gently pressing a pair of galvanometer contacts against the edge and the axle. The galvanometer needle swerved—and remained in its new position as long as the disk rotated. When he reversed the direction of spin, the galvanometer needle moved the other way. Merely by rotating between the poles of a magnet, the copper disk had become a continuous generator of electricity. A steady current—by *machine*. Here was the first dynamo, forerunner of the modern electric generator and the seed from which eventually grew the electric power industry.

But how did the current originate? The Woolwich magnet's field of influence stayed constant with time, which would seem to pre-clude induction of electricity. But its magnetic power was not uni-form in space: The greatest strength lay directly between the magnet's poles, diminishing with distance from the poles. Thus the disk's rotation continuously swept new sectors of metal from re-gions of relatively weak magnetism into regions of strong magnet-ism; from the disk's "point of view," the magnetism it encountered *was* changing, as surely as if a magnet had been thrust at it. The same was true, only in reverse, for sectors spinning away from the poles—the magnetism declining this time. The result was an in-duced current directed crosswise to the disk's spin (in other words, radially).

Faraday's dynamo was essentially Arago's wheel with a fixed,

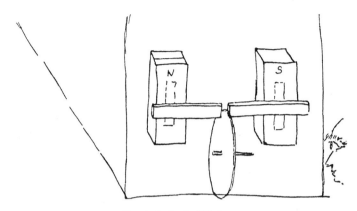

Faraday's sketch of his dynamo.

powerful magnet in place of the dangling magnetic compass needle. In both forms, an ordinary magnet induces electric current in the spinning disk. This current, like all currents, produces its own magnetism. And that magnetism, in turn, impels the compass needle to move around. In Faraday's setup, the disk's induced magnetism exerted its force on the Woolwich magnet, but that feeble force was insufficient to budge the massive pile of iron.

As intrigued as scientists were by Faraday's invention of the dynamo, they were even more impressed by this neat resolution of the mystery of Arago's wheel. Prime Minister Robert Peel stopped by the laboratory to see what all the fuss was about. When Peel inquired about the use of the new device, Faraday supposedly replied, "I know not, but I wager that one day your government will tax it." (The first successful use of an electric generator occurred in a lighthouse at Dungeness on the English Channel around 1860; a small steam engine turned a wire armature within the poles of a three-ton magnet to power the structure's carbon-arc lamp.)

Faraday tried the dynamo with disks of iron, lead, zinc, and tin. All generated electricity. Still the electricity was weak, capable of activating a sensitive galvanometer, but insufficient to even twitch a frog's leg. He improved the device with better contacts and more uniform rotation, but only to the point that it was suitable for demonstration before an audience. He was not interested in pursuing the practical side of his invention: "I have rather . . . been desirous of discovering new facts and new relations dependent on magneto-electric induction, than of exalting the force of those already obtained; being assured that the latter would find their full development hereafter." When he demonstrated his dynamo at a Friday Evening Discourse in February 1832, the frogs' legs were waiting. And this time they twitched.

8

TOIL AND PLEASURE

*The voyage of discovery lies not in seeking new horizons, but
in seeing with new eyes.*

—Marcel Proust

Faraday's initial report on electromagnetic induction was read before the Royal Society on November 24, 1831. Five days later, while vacationing in Brighton with Sarah, he wrote to Richard Phillips, brimming with satisfaction at his accomplishments: "We are here to refresh. I have been working and writing a paper & that always knocks me up in health but now I feel well again and able to pursue my subject and now I will tell you what it is about. The title will be I think *Experimental Researches in Electricity*." After a broad summary of his results so far–induction in parallel wires, induction in electrified coils, induction with magnets, the invention of the dynamo–he throws in a dig at the theoreticians' failure to explain Arago's wheel. "It is quite comfortable to me," he tells Phillips, "to find that experiment need not quail before mathematics but is quite competent to rival it in discovery and I am amused to find that what high mathematicians have announced . . . has so little foundation . . ." Faraday's Sandemanian angel finally taps him on the shoulder, and he closes, "Excuse this egotistical letter . . ." But in John Tyndall's view, Faraday had every right to be pleased, for he "overran in a single autumn this vast domain, and hardly left behind him the shred of a fact to be gathered by his successors."

Trouble began when Faraday sent word of his discovery to French physicist J. N. P. Hachette. The letter made its way to François Arago, who promptly read it at the December 26 gathering of the Academy of Sciences in Paris. Three days later the magazine *Le Lycée* published a muddled account of Faraday's work, and in a follow-up article attributed the discovery of induction to French scientists. In January 1832, two Italian researchers, Leopoldo Nobili and Vincenzo Antinori, read the first *Lycée* article, reproduced the experiments, and published their own paper on induction, taking care to properly credit Faraday for the discovery. But the journal in which their report appeared carried the date November 1831, giving the impression that the Italians, too, had preceded Faraday. (Although Faraday had announced his discovery to the Royal Society on November 24, 1831, his first paper on induction did not make it into print until early 1832.) So when William Jerden of London's *Literary Gazette* got wind of the story, he alerted readers about the great *Italian* discovery that England's own Michael Faraday had recently *repeated*.

For Faraday, it was déjà vu all over again. A decade earlier, he had been accused of plagiarizing William Hyde Wollaston's work in the electromagnetic rotations affair. Then Humphry Davy had tried to snag the glory for the liquefaction of chlorine. And now Faraday found himself robbed of credit for an even bigger discovery that was unequivocally his. He dashed off an irate letter to Jerden at the *Literary Gazette* in which he lays out the evidence of his priority, then closes with: "I never took more pains to be quite independent of other persons than in the present investigation; and I have never been more annoyed about any paper than the present by the variety of circumstances which have arisen seeming to imply that I had been anticipated." In its next issue, the *Gazette* acknowledged its error and issued an apology. Hachette also apologized to Faraday for his inadvertent role in the imbroglio, saying: "To render you complete justice, to assure you all priority, was my only aim." And to avoid any further confusion, when Nobili and Antinori's translated paper appeared in England's *Philosophical Magazine,* it carried a prominent note declaring Faraday the discoverer of electromagnetic induction. Still, Faraday learned a lesson: No longer would he release results prior to publication.

One of the prime conceptual innovations in Faraday's first report on induction was that of magnetic lines of force which, in his view, occupy space both within and around a magnet. Although nominally invisible, lines of force are made manifest when iron filings are sprinkled onto a sheet of paper placed on or slightly above the magnet—the magnetic equivalent of dusting for fingerprints. For a bar magnet, iron filings align themselves into a twin pattern of flattened ovals, curving from the magnet's north pole, through the surrounding space, into the south pole and continuing within the magnet until the oval is completed. (Actually, this is a two-dimensional "slice" through a full, three-dimensional arrangement. Also, it's best not to take the term *lines* literally; lines of force are, in most cases, actually curves.) The lines are most concentrated near the two poles of the magnet, where the magnetic power is strongest; their concentration diminishes with distance, as the magnetic power weakens.

Having defined magnetic lines of force, Faraday next established the general requirements for the onset of induction in terms of his new model. He had learned from experiment that induction requires either relative motion between a conductor and a magnet (such as a magnet being thrust into a coil) or else a change in the magnetic strength around a conductor (an electromagnet being switched on or off). In his mind's eye, he imagined a conductor moving through a region of *constant* magnetism, first parallel to the

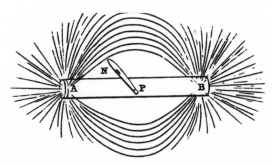

Faraday's first published drawing of magnetic lines of force ("magnetic curves," as he initially termed them).

lines of force, then across them. When moving *parallel* to the lines, following their contours like a train on a track, the conductor experiences no induction at all; current does not arise within it. When the conductor instead moves *across* the lines of force—"cutting" them in Faraday's terminology—current is induced.

Faraday also delineated the induction mechanism when a conductor moves through a region of non-uniform magnetism, as in his dynamo device. Here, he explained, the conductor is sweeping into a space with a greater or lesser concentration of magnetic lines of force, again cutting lines and generating current. Faraday didn't know the underlying process by which the electricity is produced. At this point, he used the lines of force only as a guide to predict and understand the circumstances of induction. However, he speculated that lines of forces, when cut, release magnetic energy which somehow induces electric current in the conductor doing the cutting. The convertibility of energy appealed to Faraday's belief that the diverse phenomena of nature are, in fact, manifestations of a few fundamental laws and actions. Toward the end of his career, Faraday would try in vain to convert gravity into electricity.

Faraday next applied his lines of force to induction between pairs of stationary wires or coils. Specifically, he asked himself, why does induction occur only when the primary current is changing? He envisioned the primary wire's array of magnetic lines suffusing the secondary wire. As long as the primary current stays constant, the lines of force remain stationary around the secondary wire, and no induction takes place. But if the primary current changes, the lines of force must somehow come to reflect that new condition and convey its effects into space. And it seemed reasonable that they would do so, not instantaneously, but over time. In other words, if the primary's internal wellspring of magnetic power changes, these changes might travel outward through space as *moving* lines of force.

When the current in the primary wire is switched on, Faraday reasoned, the associated magnetism builds over a brief time. This alteration in the magnetic conditions must somehow be conveyed to the secondary wire some distance away. Faraday envisioned lines of escalating magnetic force rippling outward and sweeping over the stationary secondary coil, like waves on a beach. The result would be no different if, instead, the secondary wire were to move past the lines of force, a situation where induction clearly does arise. It is the *relative* movement between the lines and the wire that somehow stimulates induction. Once the primary current reaches a steady state, the lines "freeze" into a constant arrangement, the relative movement between lines and wire disappears, and the induction process stops. That is why, in Faraday's words, "a wire at rest in the neighborhood of another carrying a powerful electric current is entirely indifferent to it . . ."

Faraday's notions that magnetic lines of force might release energy, move, and propagate magnetic effects through space hint at his closely held belief that the lines might be more than a mere visual artifice. They might be the very agents by which force is transmitted—invisible motive tentacles that impel material bodies that stray among them. And if lines of force exist for magnetism, then why not for electric charges? Or even for gravity? Perhaps the flash of lightning is but a rush of electrical lines of force. Perhaps our earth is held in orbit by the gentle embrace of the sun's gravitational lines of force. Faraday began to suspect that force arises, not when some impulse shoots instantaneously from a seat of power to a remote object—the so-called action-at-a-distance theory—but when the object encounters the lines of force that inevitably surround all magnetic, electrical, and gravitating bodies. For now, he chose to keep the broader reaches of his speculations to himself.

In his initial report on induction in 1831, and in many subsequent reports, Faraday does not specify whether the magnetic lines of force are to be taken as metaphorical or real. Nor is it clear whether the action of "cutting" a force line is meant literally. Faraday's caution was understandable in the context of his time. In entertaining such a radical notion, he stood in direct defiance of more than a century of Newtonian tradition. He was aware that his lines of force, if promoted as physical fact, would unleash a storm of opposition from the action-at-a-distance proponents. Space was considered by almost every influential scientist of the day to play no role in the transmission of the force. Space was a void whose sole function was to provide separation, to be the backdrop across which forces fly with infinite swiftness between objects. To suddenly fill space with active lines of force, à la Faraday, would have been judged heretical. And although Faraday was world renowned as an experimenter, his credentials as a theorist were marginal. In 1831, Faraday decided it would be unwise to stake his reputation on a half-formed idea. Nevertheless, proving the reality of the lines of force became a personal quest that would occupy him to various degrees for the rest of his career. As Tyndall once remarked about Faraday's unbridled tenacity, "He hated what he called 'doubtful knowledge,' and ever tended either to transfer it into the region of undoubtful knowledge, or of certain and definite 'ignorance.'"

Meanwhile, Faraday tried to induce electricity using the world's largest magnet: the earth itself. The earth's magnetic properties had been known since antiquity. Although global in extent, our planet's magnetism is relatively weak—sufficient to align a tiny compass needle or deflect incoming particles from the sun. Already in Faraday's time it was known that the earth's magnetism does not arise from magnetic rocks, but from some form of electromagnetic activity within our planet (later shown to be magnetism stemming from electrical currents in the earth's molten core).

Using only the earth's feeble magnetism, Faraday managed to generate a detectable electric current merely by flipping a large wire loop in his laboratory—cutting magnetic lines of force, in his conception. In another display of experimental prowess, he dispensed with the magnet in his Arago's wheel dynamo, and again produced electricity from the earth's magnetism alone. Faraday further theorized that global magnetism might even induce current in a stationary loop by virtue of the earth's own rotation. (He had previously shown that a spinning bar magnet induces electricity, as though cutting its own lines of force.) However, testing his assertion would be difficult: The induction effects in the opposing halves of the loop would point in the same compass direction, effectively "pushing" against each other and nullifying any tendency for current to flow. But if the circuit halves were made of different materials with different conduction properties, the induction effects would not cancel out completely; a trickle of current would remain. Faraday fashioned a 120-foot-long circuit, half copper, half iron, but the galvanometer registered nothing. Presumably, copper, iron, and other metals were too similar in electrical conductivity for any net induction to occur. He needed to find a conducting material that was very different from metal. *Water!*

Through the Royal Society, Faraday received permission to try his experiment at the large pond in Kensington Gardens. First he laid a copper wire halfway around the 480-foot-wide basin. Then he soldered each end to a copper plate, and threw these in, forming a huge circuit, part copper, part water. This time the galvanometer reacted. But after further scrutiny, Faraday attributed the minuscule current to thermal and chemical effects, not to induction. A few days later, canceling a Royal Society lecture, he repeated the experiment at the 960-foot-long Waterloo Bridge over the Thames. Again, no definitive result. At this point, Faraday envisioned an apparatus of truly epic scale: a wire stretching below the floor of the English Channel from Dover to Calais, and a return path for current through the sea itself. No, that project would be conducted, perhaps, by others in a distant future. (The predicted currents were later observed in submarine telegraph cables.)

In the third of his *Experimental Researches in Electricity,* read before the Royal Society in January 1833, Faraday addressed the nature of the various "electricities." That the plural form of the word was still in common use reflected the fact that nobody had bothered to determine whether the spark of energy from an electrostatic machine is any different from the outflow of a battery. Or whether the current induced in a coil is distinguishable from the shock of an electric eel. In his report, Faraday grapples with the physical meaning of what he is trying to measure. Others might readily accept the canonical, even intuitive, picture of electricity as a flowing fluid, imponderable or otherwise. But Faraday remains agnostic on the subject, posing a grab bag of alternatives: "By current, I mean anything progressive, whether it be a fluid of electricity, or two fluids moving in opposite directions, or merely vibrations, or, speaking still more generally, progressive forces." Through painstaking measurement with instruments of his own making, Faraday determined that static electricity, voltaic electricity, and magnetically induced electricity are indeed the same: Identical amounts of their energy rotate a galvanometer needle equally, heat a wire as much, trigger the same electrochemical reactions, and produce equivalent sensations on the tongue.

Faraday now returned to his roots in chemistry with a detailed study of electrical conduction. This work was stimulated by a puzzling observation: Liquid water conducts electricity, while frozen water—ice—does not. It had long been assumed that every substance could be classified as either an electrical conductor or an electrical insulator. The liquid and solid forms of water, for instance, have the same chemical composition and had therefore been expected to have the same electrical conductivity. Yet the conducting property of water changed as it froze or thawed. (In fact, pure water is a relatively poor conductor; most of its conductivity stems from dissolved impurities.)

Through experiment, Faraday identified an assortment of solid-liquid duos that behave like water—insulating when solid, conducting when liquid. He found that all such substances share a common property: When conducting electricity (while liquid), they simultaneously decompose—split apart—and their component elements or compounds appear at oppositely charged electrodes. Faraday's discovery challenged the still widely accepted fluid model of electricity: If electricity is indeed an imponderable fluid, coexisting but not interacting with matter, its ability to flow should not be influenced by the state of the surrounding matter.

Faraday's observations foreshadowed the modern view of conductivity. Salt (sodium chloride) is a common example of an "ionic" compound, in which positive sodium ions and negative chlorine ions are bound together by mutual electrical attraction. Solid salt is nonconducting: The sodium and chlorine ions are trapped in the crystalline matrix—"chained" in their places, as Faraday aptly put it—and will not flow under the influence of a battery. However, when heated to melting or dissolved in water, salt separates into its constituent ions. If electrodes from a battery are immersed into the liquid, the ions migrate in opposite directions. Sodium moves toward the negative electrode, where it adheres as a metal, and chlorine moves toward the positive, where it bubbles up as a gas. Such ion flow, Faraday realized, completes the electric circuit from the battery through the solution, like a liquid wire connecting the two electrodes. "Molecular" compounds, like sugar (sucrose), remain intact when dissolved, hence, their solutions are electrically neutral and nonconducting.

Fast on the heels of the conduction study, Faraday described to the Royal Society the fifth and seventh installments of his *Experimental Researches in Electricity,* about the electrical decomposition of solutions, or electrolysis. (The sixth installment concerned the role of platinum in stimulating the union of hydrogen with oxygen.) Faraday first attempted to overturn the prevailing view that the electrical force exerted by charged electrodes rips apart compounds in a solution and separates the remnants. He believed instead that the electrodes exert no force at all on the solution's contents, but serve as mere portals through which the battery's electrical influence passes into and out of the solution.

Of course, Faraday was unaware of electrons, the subatomic particles that were much later found to convey negative charge through a wire—but if he were, his published explanation of the electrolysis process might be recast as follows. The negative electrode presents a surfeit of electrons, which wrest nearby positive ions in the solution from their host molecules. These sundered molecules snatch ions from their neighbors, and the process continues like a molecular bucket-brigade throughout the solution. The result is a gradual migration of positive ions toward the negative electrode. The complementary process occurs at the positive electrode, causing movement of the solution's negative ions in the opposite direction. Together, this cross-flow of ions completes the electric circuit from the battery through the solution. And, counter to the action-at-a-distance model, all forces are intermolecular, that is, not emanating from the electrodes. (Faraday acknowledged that the force between molecules is really a Lilliputian form of action-at-a-distance; nevertheless, he was confident that such interparticle action would be explained in the future. Today we know that such forces are ordinary electrical attractions between charged particles.) To underscore the passive nature of the electrodes, Faraday applied electricity to a solution-soaked paper by zapping it with static discharges through the air. Decomposition occurred even though the electrodes—and thus their hypothesized force—were absent.

Faraday even addressed what he saw as the prejudicial aspects of language. Why retain an electrochemical vocabulary, he wondered, whose connotations reflect old ideas? For example, Faraday objected to the common use of the word *poles* instead of *electrodes*. Historically, *poles* connoted active centers of force, which was contrary to what he had just proven in his electrolysis experiments. To prompt his fellow scientists to abandon their preconceived notions, Faraday announced a series of new electrochemical terms, many of which are still in use: *electrode, anode, cathode, ion, electrolyte, electrolysis,* and others. Afterward, he told Cambridge science historian William Whewell, who had helped him craft the new language, about its frosty reception: "I had some hot objections made to them here and found myself very much in the condition of the man with his son and Ass, who tried to please every body; but when I held up the shield of your authority it was wonderful to observe how the tone of objection melted away."

From here, Faraday deduced two fundamental principles of electrolysis. First, equal amounts of electricity decompose equal amounts of substances in a solution; the electrodes' shape or size, the solution's concentration, and the battery's power are immaterial. Second, the relative mix of elements in the products of electrolysis follows a well-defined quantitative rule (which Faraday elucidated). The upshot of these principles is that both matter and electricity must come in discrete units. In terms of modern atomic theory, atoms and molecules consist of electrified particles—positive protons and negative electrons. Electricity is nothing but the flow of these particles, whether individually (electrons through a wire) or massed together (ions through a solution). When Faraday refers to a certain amount of *current* applied to a solution, he is, in fact, delineating a given number of electrons available for combination with ions. These decomposed products come out of solution in specific amounts of deposited metal or released gas.

Together, the two principles introduced in the seventh series of the *Experimental Researches in Electricity* comprise what is now known as Faraday's laws of electrolysis. The *faraday* is the agreed-upon standard unit of electric charge involved in electrolysis. (Physics maintains parallel tributes to Faraday: Here Faraday's law summarizes the ways to produce induced currents; the *farad* measures an electronic component's charge-storing capacity.) Faraday expressed the kernel of his wide-ranging work on electrolysis in an 1835 lecture: "All tends to prove that chemical affinity and electricity are but different names for the same power, and that all chemical phenomena are but exhibitions of electrical attractions." Time has proven Faraday right: The chemical bonds that link together substances are indeed manifestations of electrical forces, moderated by the distribution of electrons in the outer parts of atoms and molecules.

From the modern perspective, it may seem that Faraday erred in his view of electricity as a nonmaterial phenomenon—a wave or vibration or his intentionally vague "axis of power." However, a distinction must be drawn between *current*—the charge that actually flows through an electrical circuit—and *voltage*—the electromotive force from a battery or dynamo that drives the current. Take the gravitational analogy of a ball rolling down a hill: The ball represents the current, and the hill the voltage. On a level surface or at the bottom of the hill, the motive force of gravity—its pull—cannot impel the ball to move. But at the top of the hill, gravity is able to tug the ball downward. The positive and negative terminals of a battery create a voltage, the electrical equivalent of a hill within a circuit. The resulting electromotive force will be "felt" by electric charges everywhere within the circuit, whether they are electrons in a wire or ions in a solution. Current will flow. Faraday's "axis of power" is akin to the voltage imparted to a circuit by the battery. In later years, he would recast this motive power as a *force field,* a concept that would have profound ramifications for modern science.

By the mid-1830s, Faraday was the scientific explorer he had always wanted to be, subservient to no one except his God. He was now the Royal Institution's Fullerian Professor of Chemistry, with a no-strings-attached, 100-pound annual stipend to boost his meager salary. The Royal Society had presented him with its Copley Medal for his work on electromagnetic induction. Oxford had awarded him an honorary degree. The name *Faraday* had become synonymous with discovery. It seemed that every time he rose from his basement laboratory, he delivered yet another hard-won pearl from nature's store of mysteries. Then, barely catching his breath, he would "dive" down for more. His immersion in these depths of electricity and magnetism was so complete that he wondered whether anyone could truly comprehend his passion. "My matter . . . overflows," he wrote in 1833 to John William Lubbock, astronomer and treasurer of the Royal Society, "the doors that open before me are immeasurable. I cannot tell to what great things they may lead and I have worked neglecting every thing else for the purpose. I do not know whether Mathematical are like Experimental labours, if they are you will have an idea of my toil but at the same time of my pleasure."

9

A CAGE OF HIS OWN

If you are afraid of being lonely, don't try to be right.

—JULES RENARD

"I had begun to imagine that I thought more about Electricity and Magnetism than it was worth," Faraday confided to Cambridge historian William Whewell in 1835, "and so a notion was creeping over me that after all I was perhaps only a *bore* to my friends by the succession of papers I was bold to send forth . . ." Bore or not, the succession of papers continued.

Leaving the electrochemical realm, Faraday next turned his attention to electricity's prevoltaic roots: electrostatics. This was old-time, bread-and-butter electricity, investigated by such towering eighteenth-century figures as Henry Cavendish, Benjamin Franklin, and Charles Coulomb. Before the invention of the battery in 1800, electricity could be generated only by frictional means: A glass rod rubbed with silk acquires a positive charge, leaving the silk negative; amber become negative when rubbed with a piece of fur, while the fur turns positive. To mechanize the friction process, hand-crank electrostatic machines were invented, which unleashed spectacular, lightninglike sparks. (In my own classes, I use a motor-driven van de Graaff generator to electrify my students' hair and make it stand on end.) Touching a metal doorknob after walking across a carpet produces a palpable, less intimidating example of the flow of so-called static electricity.

In Faraday's time, the various electrostatic phenomena were explained in the same way as those of electric current: by the actions of imponderable electrical fluids. In the two-fluid model, a neutral object contains equal amounts of positive and negative electrical fluids, whereas a positively charged object has excess positive fluid and a negatively charged object excess negative fluid. Proponents of the competing one-fluid model accounted for positive and negative charges by a surplus or a dearth of the single electrical fluid. In either case, bringing objects into contact with each other supposedly incurred a transfer of excess fluid between them, with a concomitant change in each object's overall charge. Excess electrical fluid could also be siphoned off through a "ground" connection—a wire clamped to a plumbing pipe or to a rod hammered into the earth. Or, if desired, charge could be retained indefinitely by placing the object on an insulating stand. The fluid theories stood largely in accord with the observed phenomena of static electricity and had a secure mathematical basis.

In the conventional wisdom, the defining action of static charge is attraction or repulsion. That is, objects with the same type of charge—either a pair of positive objects or a pair of negative objects—repel one another. Those with unlike charge—a positive and a negative object—attract one another. The mathematical character of these electrical attractions and repulsions nestles comfortably within the bosom of Newtonian physics. Indeed, the equations of electrical force resemble a smaller-scale version of those that describe the gravitational forces linking the bodies of the solar system. In the 1830s, as in the century before, it felt good to have Newton's ghost on your side.

Predictably, Faraday's interest in electrostatics stemmed from his unease with conventional explanations of electrostatic phenomena. Electrical fluids were anathema to Faraday for two reasons: Their very existence defied experimental verification; and, in adherence to the action-at-a-distance paradigm, they attracted or repelled each other instantaneously across space. Faraday believed instead that electricity is an inherent power of matter, normally quiescent, but which can be activated in a variety of ways. Sure, the fluid theories explained electrical attraction and repulsion adequately. But why, Faraday wondered, should these two electrostatic processes alone be dubbed the gold standard of scientific proof? Why minimize the broad range of electrochemical and electromagnetic aspects of electricity, which, in Faraday's mind, were equally compelling? Faraday was convinced that there needed to be further exploration of the seemingly straightforward process of electrostatic induction—charging an electrically neutral object through proximity or contact with a charged object. (*Electrostatic* induction arises from the effect of stationary charge or the intermittent transfer of charge between objects; Faraday had previously studied induction by electric currents.)

Were electrical fluids real, Faraday argued, two observable consequences would arise: It would be possible to charge an object in an "absolute" sense, without any accompanying change in the charge state of the surroundings; and the interior space of a charged metal vessel would show signs of electrification. Faraday's plan of attack was to demonstrate the opposite: that all electrostatic phenomena are reducible to his vision of the induction process—essentially to the redistribution of charged "molecules" of ordinary matter. (He didn't know what these "molecules" were, only that they carried the property of electric charge.) First, he would prove that the induction of static charge is always a zero-sum process: Every time an object is charged, no matter what the circumstances, an equal and opposite charge will appear elsewhere—on the inducing object or machine itself, on other objects, or even on the walls of the room. Second, he would prove through meticulous measurement that the interior cavities of charged vessels are truly free of electrical charges or influences.

In his laboratory, Faraday applied charge from an electrostatic machine to various metal vessels, including a large pot borrowed from a coppersmith. (Each vessel was mounted on an insulating stand to prevent leakage of charge.) Using a sensitive charge detector—an "electrometer"—and various electrical probes that he poked around, onto, and through holes in each vessel, Faraday determined that, without exception, charge settled exclusively on each vessel's outer surface. None was found on the inner surface. Nor did he detect any sign of electrical effects within the interior cavity. And in every case, the conducting vessel itself induced an opposite charge in nearby objects in the lab. In Faraday's view, the *overall* charge in the room did not change, because the available "reservoir" of charge is fixed by the amount of matter in the room. No "absolute" charge was created independent of that which already existed in matter.

In January 1836, not content to merely probe a vessel's interior through a tiny hole, Faraday decided to dramatically increase the scale of his experiment. His "Faraday cage," as it became known, took the form of a cubical wooden frame, twelve feet on a side, sheathed in copper wire and tin foil to make the cube's "surface" a more-or-less continuous conductor. It was mounted on insulating supports in the Royal Institution's lecture hall. And there was a door so he could step inside. Perhaps he wanted to be doubly sure there are no electrical effects in the interior of a charged conductor, even when charged to many thousands of volts. Or perhaps he wanted the ultimate insider's experience of nature.

First, just to confirm that the cage had been properly constructed, Faraday placed an electrometer inside and peered from outside through a slit while the cube's surface was heavily charged by an electrostatic machine. As he had expected, the electrometer indicated that no electric charge had passed through to the interior. Then he stepped inside. "I went into the cube and lived in it, and using lighted candles, electrometers, and all other tests of electrical states, I could not find the least influence upon them, or indication of anything particular given by them, though all the time the outside of the cage was powerfully charged, and large sparks and brushes were darting off from every part of its outer surface." Again, there was no sign that any absolute charge had been created. The charge on the cage's exterior arose from induction, not from invisible electric fluids. (Automobiles and planes are Faraday cages, and can be struck by lightning with no ill effects to occupants. Boston's Science Museum uses a large Faraday cage to shield the demonstrator against deadly lightning bolts from its giant electrostatic generator.)

Faraday also explored the effects of static electricity on insulating substances. Into the gap between a pair of concentric conducting spheres, he placed different materials—shellac, sulfur, wax, glass, spermaceti. When the spheres were charged, the intervening materials—*dielectrics,* in Faraday's lexicon—became electrically charged as well. Faraday posited that the electrical force from the charged spheres induced an alignment of the dielectric's molecules, an arrangement that lodged electrical energy within the dielectric's very structure. As proof, he noted that, after the spheres had been discharged, they spontaneously recharged themselves! Evidently, electrical energy was leaking out of the dielectric back onto the spheres. Equally important, Faraday found that it took time for electricity to penetrate the dielectric. Clearly, each molecule's alignment triggered the alignment of neighboring molecules, propagating electric force through the material. The action-at-a-distance model recognized only those forces directed between the conducting spheres, and was blind to events in the intervening medium.

Faraday realized, too, that the traditional distinction between conductors and insulators was an illusion. Dielectrics conduct

The nested-sphere apparatus Faraday used to study electrostatic induction.

electricity to varying degrees, depending on their chemical makeup—or, in Faraday's new interpretation, depending on how much "strain" their molecular structure can bear in the presence of an electric force. Current flows when the applied electric force from a battery or electrostatic machine exceeds the dielectric's capacity for molecular strain. Thus, insulators can withstand considerable strain, and conductors almost none, before "breaking down" and allowing the passage of electricity. Faraday's concentric-sphere apparatus—a dielectric sandwiched between conductors—was an early form of the *capacitor,* an electrical component that is ubiquitous in modern electronics applications.

The final experiment of the series dealt another blow to the prevailing action-at-a-distance paradigm, in which all forces are conveyed in straight lines from object to object. Or as John Tyndall put it, "Gravity . . . will not turn a corner." Faraday demonstrated a case in which an *electrical* force did indeed "turn a corner." He found that he could electrify an uncharged metal sphere by the approach of a charged sphere, even when there was an obstacle between them. The electrical influence of the charged sphere must have followed a curved trajectory to reach its uncharged sibling. The intervening space was therefore not passive, as action-at-a-distance proponents would have proclaimed. It played an active role in the transmission of the force. In Faraday's view, the obstacle was bypassed because the charged sphere's electrical lines of force bent around the obstacle and converged on the uncharged sphere.

Faraday read the eleventh of his *Experimental Researches in Electricity* to the Royal Society in November and December 1837. His opening sentence reflects his precarious position, as he pleads with the audience to keep an open mind. "The science of electricity is in that state in which every part of it requires experimental investigation; not merely for the discovery of new effects, but what is just now of far more importance, the development of the means by which the old effects are produced, and the consequent more accurate determination of the first principles of action of the most extraordinary and universal power in nature: and to those philosophers who pursue the inquiry zealously yet cautiously, combining experiment with analogy, suspicious of their preconceived notions, paying more respect to a fact than a theory, not too hasty to generalize, and above all things, willing at every step to cross-examine their own opinions, both by reasoning and experiment, no branch of knowledge can afford so fine and ready a field for discovery as this."

For two years, Faraday had assembled his array of evidence against electrical fluids and the action-at-a-distance theory. In this effort he was almost alone among scientists. "[I]n whatever way I view it," he told his colleagues, "and with great suspicion of the influence of favourite notions over myself, I cannot perceive how the ordinary theory applied to explain induction can be a correct representation of that great natural principle of electrical action." He knew that his ideas marked a frontal assault on long-established theories posed by such notables as Charles Coulomb, André-Marie Ampère, and Siméon Denis Poisson. He was trying to overturn the belief system in a branch of physics—electrostatics—that was firmly supported by mathematical theories he himself could not understand. He had previously challenged action-at-a-distance in the area of electrochemistry, but that field's theoretical underpinnings had not yet crystallized; there his conclusions were welcomed because they imposed order on what had been chaos.

While honored in public, Faraday was scorned by many of his university-trained counterparts, who found both his manner of scientific expression obtuse and his lack of mathematical rigor frustrating. Some of them probably looked on bemused as the self-taught, experimental genius struggled to make himself understood in the theoretical arena. Faraday was aware of the disadvantage under which he operated. To Ampère, he once confided: "I am unfortunate in a want of mathematical knowledge and the power of entering with facility into abstract reasoning. I am obliged to feel my way by facts placed closely together, so that it often happens I am left behind in the progress of a branch of science not merely from the want of attention but from the incapability I lay under of following it, notwithstanding all my exertions . . . I fancy the habit I got into of attending too closely to experiment has somewhat fettered my powers of reasoning, and chains me down, and I cannot help now and then comparing myself to a timid ignorant navigator who (though he might boldly and safely steer across a bay or an ocean by the aid of a compass which in its actions and principles is infallible) is afraid to leave sight of the shore because he understands not the power of the instrument that is to guide him." Yet Faraday was also quick to point out that the mathematical approach has its own pitfalls when stacked against experiment: "I have far more confidence in the one man who works mentally and bodily at a matter than in the six who merely talk about it . . . Nothing is so good as an experiment which whilst it sets error right gives us a reward for our humility in being refreshed by an absolute advancement in knowledge."

The first volume of Faraday's *Experimental Researches in Electricity* was completed in 1838 with his fourteenth paper in the series. By the time the entire work was done in 1855, there were three volumes of 1,114 pages, comprising twenty-nine research reports and a host of other papers. The paragraphs are numbered from 1 to 3362, with frequent cross-references, an indexing scheme Faraday may have used to aid his troublesome memory. The level of detail is staggering. Here is a complex narrative that mingles success with failure, revelation with mystery. The reader becomes an insider, peering over Faraday's shoulder in an uncensored view of scientific investigation. Upon receiving a copy of *Experimental Researches,* German physicist Peter Riess told Faraday, "If Newton, not without reason, has been compared to a man who ascends to the top of a building by the help of a ladder, and cuts away most of the steps after he has done with them, it must be said that you have left to the follower, with scrupulous fidelity, the ladder in the same state as you have made use of it."

In reviewing the massive work, John Tyndall commented, "The salient quality of Faraday's scientific character reveals itself from beginning to end of these volumes; a union of ardour and patience—the one prompting the attack, the other holding him on to it, till defeat was final or victory assured. Certainty in one sense or the other was necessary to his peace of mind. The right method of investigation is perhaps incommunicable; it depends on the individual rather than on the system, and the mark is missed when Faraday's researches are pointed to as merely illustrative of the power of the inductive philosophy. The brain may be filled with that philosophy; but without the energy and insight which this man possessed, and which with him were personal and distinctive, we should never rise to the level of his achievements. His power is that of individual genius, rather than of philosophic method; the energy of a strong soul expressing itself after its own fashion, and acknowledging no mediator between it and Nature."

As he approached his fiftieth year in the late 1830s, Faraday's frantic work pace exacted an increasing toll on his physical and emotional well-being. He began to complain of confusion and giddiness. The "nervous headaches" that had intermittently plagued him since youth were now frequent and disabling, sometimes lasting weeks, even months. Memory loss, which had arisen in his twenties, now became "so treacherous," he confided to his overseas friend Christian Friedrich Schoenbein, that he could not "remember the beginning of a sentence to the end." Trips to the countryside restored him briefly, then the symptoms returned. Physicians diagnosed him vaguely with "mental fatigue." But Faraday feared worse: a chronic and debilitating affliction that might end his career.

To preserve his flagging energy, Faraday retreated into a circumscribed universe anchored in research, family, the Sandemanian church, and the Royal Institution. He spurned office of any kind. Professional reading was confined to his specialized interests, as he once related to Astronomer Royal George Biddell Airy: "I have been convinced by long experience that if I wish to be respectable as a scientific man it must be by devoting myself to the unremitting pursuit of one or two branches only; making up in industry what is wanting in force."

Social invitations were routinely rejected, except those from the presidents of the Royal Institution or the Royal Society. All others got a polite brush-off. To London antiquary John Britton, Faraday once wrote: "I am greatly obliged to you for your invitation but am unable to accept it. I am reluctantly forced to forgo my friends company at table from January to June [when the Friday Discourses are held] and very rarely at other times dine out. In fact I am not a social man." To civil engineer John Rennie: ". . . am under the necessity of declining it because of a general rule which I may not depart from without offending many kind friends . . ." To actor and theater manager William Charles Macready: ". . . now several years since I have dined or gone out and I cannot break through my rule (which is a very strict one) . . ." To Royal Academy painter Thomas Phillipps: ". . . am so circumstanced that I cannot possibly avail myself of your favour." Again to Astronomer Royal George Biddell Airy: ". . . quite shut out from the enjoyment by the circumstance of the day being lecture day with me here & full of occupation." And to philanthropist and heiress Angela Georgina Burdett Coutts, who had hoped to socialize with Faraday before a public gathering at the Adelaide Gallery of Practical Science: "Mr. Faraday is very grateful to Miss Coutts for the honor of her invitation and deeply regrets that he cannot accept it . . . Mr. Faraday hopes however to offer his respects to Miss Coutts at a later hour in the Evening by the side of the Electric Eel."

Visitors to Faraday's laboratory were turned away as well. "I called on Faraday this morning," physicist Charles Wheatstone related to a colleague, "and was told that this was one of the days on which he denies himself to every body for the purpose of pursuing uninterruptedly his own researches. He will be visible tomorrow." Even longtime friend Benjamin Abbott had trouble getting in. Laboratory assistant William Fletcher Barrett recalled that Abbott had once "called to see Mr. Faraday, whom he had not met for many years. On inquiring at the Royal Institution, the hall-porter told him that Mr. Faraday was at work, and could not be seen by anyone, indeed, that he dare not even take a visitor's card down to him. On Mr. Abbott's explaining that he was an old friend, the hall-porter suggested . . . that, as it was near one o'clock . . . he might catch sight of Mr. Faraday coming upstairs."

On several occasions, Sarah Faraday approached her husband about moving outside the Royal Institution so that he might avoid the "continual calls upon his time and thought; but he always shrank from the notion of living away from the R. I." "I have been here so long," Faraday mused, ". . . that I feel as if I were a limpet on a rock and that any chance which might knock me from my position would leave me but little hopes of attaching myself anywhere again. So much for habit which is just as strong in matters of feeling as in matters of body."

On Friday evening, November 29, 1839, Dr. Peter Latham, Physician Extraordinary to the Queen, made an emergency visit to the Faradays' apartment. Faraday looked pale, had strabismus of one eye, and was so dizzy he could barely walk. Dr. Latham bled Faraday at both temples, then advised rest. "He confessed to me," read Latham's clinical note, "that his mind has for a long time been dreadfully overworked—that, besides his own abstruse speculations upon electricity, he has had lectures pressed upon him, and is continually consulted about all sorts of subjects, which people fancy are quite easy to him, but which require considerable thought." When Faraday delivered a lecture the next day, Latham warned Faraday's colleague William Brande: "He looks up to this work; but in truth he is not fit and if he is pressed he will suddenly break down."

Upon Latham's advice, Faraday took the month off in Brighton. Other respites followed, and visits to the laboratory became intermittent. "My medical friends have required me to lie by for a twelve month and give me hope that memory (without which it is hard work to go on) may perhaps come on. They want to persuade me that I am mentally fatigued and I have no objection to think so. My own notion is, I am permanently worse: we shall see." On September 14, 1840, Michael Faraday walked out of his laboratory at the Royal Institution, not knowing when—or whether—he would return.

10

AN EXCELLENT
DAY'S WORK

Every wall is a door.

—Ralph Waldo Emerson

Sometime in the early 1840s, Faraday grabbed a scrap of paper and scrawled, "[T]his is to declare in the present instance, when I say that I am not able to bear much talking, it means really, and without any mistake, or equivocation, or oblique meaning, or implication, or subterfuge, or omission, that I am not able; being at present rather weak in the head, and able to work no more."

After the acute attack in 1839 that forced him to scale back and ultimately suspend his research, Faraday sank into a period of melancholy and self-enforced isolation. "You know that I am a recluse & unsocial," he wrote to French scientist Jean-Baptiste-André Dumas in 1840, "and have no right to share in the mutual good feeling of Society at large for the man that does not take his share of goodwill into the common stock has no claim on others. Such is not the case I hope from any cold or morose feeling in the heart but from particular circumstances amongst which are especially mental fatigue and loss of memory." Faraday complained to his friend

Christian Friedrich Schoenbein of "low nervous attacks" and "memory so treacherous that I cannot remember the beginning of a sentence to the end." His hand became "disobedient to the will," making the simple act of writing a letter a challenge.

As if his manifest symptoms were not distressing enough, there was the gnawing uncertainty about his long-term prognosis. Might his affliction abate? Might he someday pursue science again with restored vigor? Or was his mental stamina permanently destroyed? His condition was unpredictable. A good day might be followed by a bad one, hopefulness extinguished by depression, only to see his spirits rise again when his mental fog lifted. He constantly monitored his physical, cognitive, and emotional state, searching for the faintest glimmer of a return to normalcy. The only cure, physicians and friends told him, was complete rest. Or in practical terms, abandonment of his laboratory and his investigations of nature–a cruel reversal of his self-coined motto, "Work, finish, publish."

Sarah Faraday strove to counter her husband's predilection toward overwork. While in London, the circus and the Zoological Gardens became regular stops, often with a relative in tow. Extended retreats to Brighton, Margate, and other seaside resorts were routine. Here daytime wanderings were followed by family-centered evenings of music, word games, charades, and readings from the Bible, Shakespeare, or Byron. Sarah quashed an invitation from Andrew Crosse, a wealthy amateur scientist, to vacation at his estate. "Mrs. Faraday drew me aside," Crosse's wife Cornelia recalled, "and candidly told me in much kindness, and with true wifely wisdom, that our house, was of all places, the one where she could not permit her husband to spend his holiday. She was well aware that Fyne Court had all its laboratories and foundries, in short had electrical arrangements from garret to basement, and she foresaw that Faraday, instead of resting his brains, would be talking science all day long."

During the summer of 1841, the Faradays spent three months

touring Switzerland and Germany, accompanied by Sarah's brother George Barnard, an artist, and his wife Emma. While the two women sauntered through quaint villages, the men headed into the mountains on all-day "sketching excursions," Barnard with his pad and pencils and Faraday with a book and spyglass. From Switzerland, Sarah wrote to her husband's longtime friend Edward Magrath, formerly of the City Philosophical Society, now secretary of the Athenaeum Club: "Mr Faraday seems very unwilling to write letters he says it is quite a labour to him, and that every one advises that he should take thorough rest and that he is quite inclined to do so. I can certainly say nothing against all this but I am anxious that such an old friend as you are should not be neglected altogether . . ." In the same letter, written six weeks into their trip, she writes that her husband "enjoys the country exceedingly and though at first he lamented on absence from home and friends very much, he seems now to be reconciled to it as a means of improving his general health, his strength is however very good he thinks nothing of walking 30 miles in a day (and very rough walking it is you know) and one day he walked 45 [miles from Leukerbad to Thun, over the Gemmi pass, in just over ten hours] which I protested against his doing again tho' he was very little the worse for it, I think it is too much . . . but the grand thing is rest & relaxation of the mind which he is really taking . . ."

Faraday appended his own letter to his wife's: "Now as to the main point of this trip i.e. the mental idleness you can scarcely imagine how well I take to it and what a luxury it is, the only fear I have is that when I return friends will begin to think that I shall overshoot the mark; for feeling that any such exertion is a strain upon that faculty, which I cannot hide from myself is getting weaker, namely memory, and feeling that the less exertion I make to use that the better I am in health & head, so my desire is to remain indolent mentally speaking and to retreat from a position which should only be held by one who has the *power* as well as the will to be active. All this however may be left to clear itself up as the time proceeds . . ." Privately, Faraday noted, "I would gladly give half my strength for as much memory, but—what have I to do with that? Be thankful."

Faraday returned intermittently to the scientific scene in the early 1840s. He delivered several Friday Evening Discourses, continued his consultations on lighthouses, and (reluctantly) led government inquiries into explosions at a gunpowder factory in Essex and a coal mine in Durham. He did return to the laboratory in June 1842 to ascertain why steam venting from a boiler is electrically charged. The description of his experiment and his conclusion (that the charge arises through friction between steam-borne water droplets and the orifice through which the steam issues) formed the eighteenth in his series of *Experimental Researches in Electricity* in 1843. A lengthier but equally workmanlike study from May 1844 to August 1845 involved the liquefaction of gases. Despite this activity, the deterioration of his faculties was never far from his mind.

Ada Lovelace, high-strung daughter of Lord Byron, was in awe of Faraday and, in 1844, begged to become his laboratory assistant—or, in her words, a bride to science. Faraday took some pleasure in the attention, but rejected her offer. "You have all the confidence of unbaulked health & youth both in body & mind," he wrote. "I am a labourer of many years' standing made daily to feel my wearing out. You, with increasing acquisitions of knowledge, enlarge your views and intentions; I, though I may gain from day to day some little maturity of thought, feel the decay of powers, and am curtailing to a continuing process of lessening my intentions and contracting my pursuits . . . You do not know and should not know but that I have no concealment on this point from you, how often I have to go to my medical friend to speak of giddiness and reeling of the head &c, and how often he has to bid me cease from restless thoughts and mental occupation and retire to the seaside and inaction." (Faraday might also have shied away from the fire of

Ada's emotion. "I have long been vowed to the Temple," Ada explained to him, "the Temple of Truth, Nature, Science! And every year I take vows more strict, till now I am just entering those portals & those mysteries which cut of[f] all retreat, & bind my very life & soul to unwearied & undivided science at its altars henceforward. I hope to die the High-Priestess of God's works as manifested on this earth, & earn a right to bequeath to my posterity the following motto, *'Dei Naturesque Interpres'* [*Interpreter of God and Nature*].")

In late 1844, Faraday published the second volume of his *Experimental Researches in Electricity,* collecting his fifteenth through eighteenth series of studies plus some thirty other papers on electricity. He was by now resigned to the likelihood that it might be his last significant contribution to science. His exhaustion was more than mere depletion of energy; it was a guttering of the flame of insight that had previously guided his exploration of nature. During the 1830s, he had blazed a largely solitary trail through the essential behaviors of electricity and magnetism and had established a fundamental link between these once-diverse phenomena. As fruitful as these studies were, his underlying goal had always been to unite his observations into an all-encompassing picture of force, matter, and light. Yet the means to establish this synthesis eluded him. Thus Faraday's explorations had ceased for want of direction, as well as for want of strength. Then, in August 1845, a letter arrived from a young researcher whom he had recently met. It proved to be the compass Faraday needed to guide him back onto the path of discovery.

William Thomson, the future Lord Kelvin, had been a nineteen-year-old mathematics student at Cambridge when he first encountered Michael Faraday's *Experimental Researches in Electricity.* He had pored through them, appalled to find not a single equation or mathematical analysis among the dense prose. In March 1843, Thomson jotted in his diary, "I have been reading . . . of Faraday's researches. I have been very much disgusted with his way of *speaking* of the phenomena, for his theory can be called nothing else." However, after further study, Thomson's disgust had changed to admiration. By the time he introduced himself to England's great experimentalist in June 1845 at the annual meeting of the British Association for the Advancement of Science, he had become, in his own words, "inoculated with Faraday fire." After the meeting, Thomson applied his own analytical prowess to Faraday's theorized lines of force. To his surprise, he found the half-baked concept to be sound, if not quite convincing. (In a mathematical sleight-of-hand, Thomson recast the standard equations of heat conduction to derive the properties of force lines around an electrically charged object.) As the summer progressed, Thomson realized that his mathematical renderings of the lines of force hinted at effects that might be observable in the laboratory. Most significantly, he informed Faraday on August 6, 1845, that electricity or magnetism should affect the passage of light through a transparent object. Faraday responded immediately: He had already sought such an effect, without success, more than a decade earlier. But reading now that a "mathematical" scientist—even one so junior in rank as Thomson—saw promise in his ideas motivated him to try again. "I purpose resuming this subject hereafter," he wrote to Thomson with palpable excitement. Faraday cast aside all other work to focus on the new project: the relationship of electricity and magnetism to light.

On August 30, 1845, Faraday returned to his laboratory with a renewed sense of mission. He would use a modified form of light—polarized light—to assess whether a luminous beam is influenced by electric or magnetic forces. Light rays possess a vibratory quality that is transverse to the direction along which they propagate. If these transverse oscillations fall within the same plane—horizontal, vertical, or otherwise—the light is said to be polarized. By analogy, shaking a taut rope repeatedly up and down creates "vertically polarized" oscillations; side-to-side shaking produces oscillations that are "horizontally polarized." Incandescent bulbs, candles, and the

William Thomson.

sun emit light that is effectively unpolarized; oscillations occur in many different planes simultaneously. Sunshine reflecting off a flat, horizontal surface–a pond, a field, a parking lot–is polarized mostly in a horizontal plane; to filter out this glare, sunglasses incorporate vertically polarized lenses. (Likewise, a tossed stick will more likely pass between vertical slats of a fence if oriented vertically, not horizontally.) In Faraday's time, polarized light was already being used to probe the internal structure of crystals, glass, and various liquids.

Faraday obtained polarized light by reflecting the glow of an oil lamp (from his lighthouse consultations) off an upright pane of glass. This beam he directed through the transparent test substance and subsequently through a polarized lens. He next rotated the lens until the lamp's image was extinguished, then applied electricity or magnetism to the transparent substance. Any change in the "blackout" condition would indicate a connection between electrical or magnetic force and light; that is, the force would have altered the lamplight's original polarization, rendering the lamp once again visible through the lens.

In the first set of trials, Faraday passed polarized light through electrified solutions in a glass trough–distilled water, dissolved sugar, sulfuric acid, copper sulfate, turpentine. He sent an electric current first parallel to the light beam, then across the beam. He then reversed the current's direction. He tried weak current, strong current, pulsed current, current rising suddenly, current rising gradually, current from a battery, current from an induction coil, sparks. He varied the size and spacing of the electrodes and at one point tested the induction coil's "terribly powerful" jolt on his body to make sure it was working. He continued with other transparent substances: air, plate glass, quartz, rock crystal, Iceland spar. His laboratory diary chronicles in minute detail four days of unremitting effort. And repeated failure: "no effect"; "still no effect"; "results all negative"; "no sign of change"; "no difference"; "all was nul."

Having failed to detect any influence of electricity on light, Faraday shifted to magnetism. He placed the poles of an electromagnet near the light path, then directed the polarized beam, in turn, through a variety of transparent substances. Once again, he rotated the lens until the image of the lamp disappeared. And once again, regardless of any adjustments to the magnetic setup–reversing the magnet's polarity, varying its strength, altering the pole positions–Faraday saw no sign of the lamp through the lens. Nor does his laboratory diary shed any light on his state of mind, as he stoically chronicles repeated failure. Paragraph after paragraph, page after page, nothing but mind-numbing particulars, penned with drab uniformity in his own hand. Until September 13, 1845, paragraph 7,504. Here appears, in stout capital letters and underlined three times, a large exclamatory "BUT." That single word, an island rising above a tedious sea of ink, illuminates Faraday's joy as surely as the lamplight that suddenly illuminated his eye. He could now see the image of the lamp cast by the lens. This was the proof he had been seeking. The electromagnet had rotated the beam's polarization, rendering the lamp visible through the lens. When he cycled the power to the magnet off and on, the lamp disappeared and reappeared on cue. Magnetism does influence light. Faraday concludes the discovery paragraph in his diary with characteristic understatement: "This fact will most likely prove exceedingly fertile and of great value in the investigation of . . . conditions of natural force."

Ironically, the key to Faraday's success lay in his much-detested optical glass project from the 1820s. Among its products were test slabs of highly refractive lead borate glass. Such "heavy glass" affects light more than ordinary glass and, Faraday suspected, might therefore enhance the sorts of subtle optical effects he was looking for. Which it did.

Faraday found that the beam's polarization rotated slightly whenever a pair of unlike magnetic poles were placed on the same side of the glass slab. (In this arrangement, the light passes nearly parallel to the lines of magnetic force.) And just to be sure that it was magnetism affecting the light and not some physical distortion of the glass itself, Faraday shut off the magnet and squeezed the glass with his hand. No effect was seen. Satisfied with his work, Faraday concludes the September 13 diary entry: "Have got enough for today."

On September 18, 1845, Faraday repeated the experiment with a stronger electromagnet. He devotes twelve pages of his diary to dozens of trials with different substances. Now his entries are laced with phrases like "effect was good"; "effect was best"; "very fine effect"; "effect was greatly increased." He closes the September 18 record: "An excellent day's work."

In subsequent experiments, Faraday determined that magnetism does not act *directly* on light, but requires a mediating substance to convey its power. The nature of the effect—a rotation of polarized light—was the same, regardless of the very different molecular structures of the transmitting substances or whether they were liquid or solid. Thus, Faraday claimed, the experiment proves that light and magnetic force are related, even though matter must intervene: "I believe that, in the experiments I describe in the paper, light has been magnetically affected, *i.e.* that that which is magnetic in the forces of matter has been affected, and in turn has affected that which is truly magnetic in the force of light . . ." He also pointed out that none of the transmitting substances became magnetized in the ordinary sense, as a bar of iron would have. Here was a new state of magnetism—*diamagnetism,* he called it—arising, not within magnetic iron or lodestone, but within diverse examples of *common* matter. Nor, Faraday suspected, was this diamagnetic state restricted to transparent bodies: Wood, stone, and metal should exhibit the property as well. And he set out to prove it.

The headaches and memory loss that had nearly ended his career were now an accepted, if inconvenient, part of Faraday's working life. They were the price of productivity, a price he willingly paid to again experience the thrill of discovery. "At present," Faraday wrote Schoenbein from Brighton in November 1845, "I have scarcely a moment to spare for any thing but work. I happen to have discovered a direct relation between magnetism & light also Electricity & light—and the field it opens is so large & I think rich that I naturally wish to look at it first . . . I actually have not time to tell you what the thing is—for I now see no one & do no thing but just work. My head became giddy & I have therefore come to this place but still I bring work with me. When I catch time I will tell you more." And to Auguste de la Rive, a month later: "I have had your letter by me on my desk for several weeks intending to answer it but absolutely I have not been able for of late I have shut myself up in my laboratory and wrought to the exclusion of every thing else . . . I am still so involved in discovery that I have hardly time for my meals & am here at Brighton both to refresh & work my head at once and I feel that unless I had been here & been careful I could not have continued my labours."

After reporting his findings to the Royal Society, Faraday turned his attention from magnetism's effect on light to its effect on matter, specifically, the diamagnetic state. Magnetism had long struck him as an anomaly among nature's forces. He could not understand a divine plan in which electricity and gravity are ubiquitous in matter, yet magnetism is exhibited by a mere handful of materials. Perhaps the key to righting this lopsided arrangement of forces lay in the newly discovered phenomenon of diamagnetism.

To ascertain whether all substances are inherently magnetic, Faraday placed various material samples in a tiny sling of writing paper, which he tied to a thread of cocoon silk and suspended between the poles of a magnet. Glass, water, alcohol, ether, iodine, olive oil, sugar, wood, ivory, caffeine, sealing wax, mutton, beef (both fresh and dried), blood, leather, apple, bread—some fifty substances in all. Every substance aligned itself oddly *crosswise* to the imposed magnetic force. (The alignment, perpendicular to a compass needle's, arises from the action of magnetism on an atom's orbiting electrons; diamagnetic objects are *repelled* by magnetic poles.) Faraday remarked, "It is curious to see such a list as this of bodies presenting on a sudden this remarkable property, and it is strange to find a piece of wood, or beef, or apple, obedient to or repelled by a magnet. If a man could be suspended, with sufficient delicacy . . . and placed in the magnetic field, he would point equatorially [that is, perpendicular to the lines of force]; for all the substances of which he is formed, including the blood, possess this property." In other words, every material possesses an innate magnetic character to a greater or lesser degree, with iron, nickel, and cobalt being at the strong end of the magnetic spectrum and substances such as glass and wood at the weak end. Just as Faraday had believed, magnetism truly is a universal force, on par with electricity and gravity. The once-disparate pieces of nature's great puzzle were coming together before his eyes.

11

NOTHING IS TOO WONDERFUL TO BE TRUE

The altar cloth of one aeon is the doormat of the next.

—Mark Twain

Cheering the tumblers at the circus; plying the surf off Brighton; ascending alpine passes in Switzerland—these exertions, nominally curative of Faraday's ills, were powerless to divert him from pondering nature. Regularly removed from his work space by fatigue or headaches, Faraday nurtured the laboratory of his mind—a test bed of the imagination, churning with ideas on force, light, and matter. Swirling around in his head were heretical ideas, so far afield of the norm that he had been cautious about openly promoting them. His drawerful of medals didn't matter to his critics; nor did the raft of less ornamental honors, or even his international fame. He would always be the outsider, simultaneously praised and disparaged by university-trained practitioners—son of a blacksmith, and now himself a kind of "scientist-smith" in their eyes, indisputably good at forging experiments, but ill-equipped to tackle theoretical problems. (It didn't help either that he refused all higher office; turning down the presidency of the Royal Society in 1857, he remarked to Tyndall, "I must remain plain Michael Faraday to the end.") Until now, Faraday had advanced his speculations in the spirit of scientific inquiry: They were but hypotheses, he would remind everybody. Mere suggestions. Alternatives to be explored. That his beliefs about nature differed, sometimes radically, from the conventional scientific wisdom could not be helped. He had always been guided by his bedrock devotion to the observable: theory anchored in fact, and fact anchored in well-wrought experiment. By the mid-1840s, his adherence to this dictum resulted in some of his wildest speculations yet.

Fortified now by age and experience, Faraday intensified his assault on the scientific edifice. His first target: the reigning model of the atom. Faraday was not denying the existence of the atom per se, only its materiality. "The atomic doctrine . . ." he cautioned, "is not so carefully distinguished from the facts . . . though it is at best but an assumption." On January 19, 1844, at a Friday Evening Discourse, titled "A speculation touching Electric Conduction and the Nature of Matter," Faraday asked his audience to consider what is truly known about the atom. No one had ever seen an atom nor cast any experimental illumination on its nature. How, therefore, could one justify the accepted picture of the atom as a tiny particle unaccountably endowed with the means to generate force? "[A]ll our perception and knowledge of the atom, and even our fancy," Faraday asserted, "is limited to ideas of its powers: what thought remains on which to hang the imagination of [a particle] independent of the acknowledged forces? . . . Now the powers we know and recognize in every phenomenon of the creation, the abstract matter in none; why then assume the existence of that of which we are ignorant, which we cannot conceive, and for which there is no philosophical necessity."

While atoms themselves lie beyond our ken, Faraday reminded the audience, forces are clearly perceptible. The tug of a magnet, the fall of a raindrop, the electrochemical sundering of a solution.

To the critical explorer of nature, therefore, the notion of the "particle-atom" must be abandoned in favor of what alone is demonstrably real: lines of force. At the heart of every atom, Faraday held, is a point devoid of physical substance, yet unequivocally defined by the convergence of lines of force. Thus an atom is not a tiny ball of matter distinct from its surrounding space; it is, in fact, a vast structure: a massless seat of power *plus* its associated force, whose lines impinge on those of every other atom across the room, across the solar system, across the universe. As Faraday later wrote, "[T]hat which represents size may be considered as extending to any distance to which the lines of force of the particle extend: the particle indeed is supposed to exist only by these forces, and where they are it is." In the Wonderland of his imagination, Faraday's atom played the Cheshire cat—the lines of force remain, while the material body vanishes. (The modern conception of subatomic particles embodies some of the strangeness intuited by Faraday. The fundamental units of matter are not like tiny marbles; they are phantoms that defy direct examination, forever flitting about in a cartoon-cloud of insubstantiality, while imparting to the surroundings all the requisite effects of mass, gravity, and electric charge.)

Space, Faraday believed, is suffused with a web of tensioned force lines, exerting their influence through contact with each other. Such interactions could account for all the observed mechanical and chemical properties of matter without resorting to ad hoc powers of tiny particles. In space, not in the atom, lay the active agents of electricity, magnetism, and gravity: the lines of force. "The powers around the centres," he wrote, "give these centres the properties of atoms of matter; and these powers again, when many centres by their conjoint forces are grouped into a mass, give to every part of that mass the properties of matter."

To support his contention, Faraday offered a paradox regarding the nature of space and matter, specifically, the distinction between an electrical insulator and an electrical conductor. Suppose standard atomic theory holds, and matter consists of tiny particles. Each particle is separated from its neighbors by an intervening space. (If atoms abut, he argued, matter would not contract as it does under pressure or cold. In fact, he used chemical data to show that the spacing of atoms must be substantially larger than atoms themselves.) This interatomic void is the only continuous path through which electricity can flow. Thus, the space within an electrical insulator must also be *insulating;* otherwise, electricity would flow freely through such an object. By the same token, the space within an electrical conductor must be *conducting* for electricity to flow within. How can space be both an insulator and a conductor of electric force? As Tyndall fairly put it, Faraday "tosses the atomic theory from horn to horn of his dilemmas."

Faraday provided his own solution to the conundrum: electrical lines of force. The lines span the gap between atoms, providing a potentially conducting highway through an object—"an atmosphere of force," he called it. To Faraday, there was no sharp distinction between insulators and conductors: All materials are conductive to some degree. The tendency of a substance to conduct electricity is reflected in the structure of its internal lines of force. Indeed, he reminded the audience, it is the lines of force that endow every substance with its unique chemical and physical properties.

Faraday closed the Discourse with a plea to his colleagues to give his revolutionary ideas a fair hearing: "I cannot doubt but that he who, as a wise philosopher, has most power of penetrating the secrets of nature, and by guessing by hypothesis at her mode of working, will also be most careful, for his own safe progress and that of others, to distinguish that knowledge which consists of assumption, by which I mean theory and hypothesis, from that which is the knowledge of facts and laws; never raising the former to the dignity or authority of the latter, nor confusing the latter more than is inevitable with the former."

It would be almost two years before anyone in the scientific community awoke to the merit of Faraday's speculation. In November 1845, after inspiring Faraday to experimentally verify the connection between magnetism and light, William Thomson published the first mathematical treatment of electrical and magnetic lines of force. His conclusion: "All the views which Faraday has brought forward and illustrated or demonstrated by experiment, lead to this method of establishing the mathematical theory, and, as far as the analysis is concerned, it would, in the *general* propositions, be even more simple, if possible, than [the currently accepted theory of static electricity] . . ." Thomson did not prove that lines of force are real, only that their mathematical avatars effectively represent the observed action of static electricity and magnetism. There was still only one person in the world convinced of the reality of the lines of force. Faraday himself.

On April 3, 1846, Charles Wheatstone burst out of the Royal Institution and bustled down Albemarle Street like a flushed grouse. It was mere minutes before his Friday Evening Discourse was to begin and, if the oft-told story is true, the sight of the notorious heckler, Joseph Crabtree, in the audience had spooked the famously skittish Wheatstone. (His pathological shyness was long-standing; even as a child he preferred the solitude of the attic to the company of schoolmates.) At nine o'clock, Michael Faraday stepped to the front of the lecture hall, made apologies, and related to the audience what little he knew about his absent colleague's latest invention, an electromagnetic stopwatch. With the rest of the hour to fill, Faraday launched into an impromptu monologue, verbalizing some half-formed thoughts about a possible relation between light and lines of force. According to John Tyndall, the gathered listeners were treated to "one of the most singular speculations that ever emanated from a scientific mind."

Nearly fifty years had passed since another Royal Institution scientist, Thomas Young, had found clear evidence that light exhibits characteristics of a wave, contrary to the then-accepted model of light as a particle. Although there remained lingering debate as to whether light is an undulation or a fleet grain of some sort, wave supporters—Faraday among them—were ascendant by the mid-1800s. Yet there was a glaring flaw in the wave model, at least in Faraday's mind: All known waves required a *medium* through which to convey their energy from one place to another. A ripple spreading over a pond involves the oscillatory movement of water. Sound propagates as pressure waves in air or some other conducting substance. So how could light waves traverse the vacuum of space? What was the medium that conveyed ripples of luminous energy from a star to the human eye? The passage of light through space, argued wave proponents, indicates that a vacuum is, in fact, not devoid of all essence. There must exist a substance—the *luminiferous ether,* they dubbed it—that permeated space and supported the mechanical undulations of light waves.

Extraordinary stuff, this ether. It could not be seen, no matter how hard one tried. It was perfectly elastic, bending effortlessly to the will of light waves. It offered no resistance to the passage of light or to planets, stars, or other material objects hurtling through it. (If it did, the earth would slow and spiral into the sun.) The luminiferous ether was truly a substance without substance. To Faraday, who had no use for imponderable fluids (his mentor, Humphry Davy, too, had railed against them decades earlier), the existence of such a pervasive, undetectable, gravity-free essence in space was intolerable. On this point, history was on his side. Electricity and magnetism, once believed to be imponderable fluids, were increasingly regarded as ordinary properties of matter. And heat, also a reformed imponderable, was seen as energy arising from the incessant motions of atoms and molecules. There was simply no place for the luminiferous ether in Faraday's philosophy. But if the ether were banished, then what mechanism transmits radiant energy through the vacuum of space? "The view which I am so bold as to put forth," Faraday told his Royal Institution audience, "considers . . . radiation as a high species of vibration in the lines of force which are known to connect particles and also masses of matter together. It endeavours to dismiss the aether, but not the vibration." On that April evening in 1846, with Charles Wheatstone in full flight through the streets of London, Faraday laid out the basis of the modern electromagnetic theory of light.

Faraday recorded his radical idea in a letter to *Philosophical Magazine,* printed under the title, "Thoughts on Ray-vibrations." "[F]rom first to last," he alerts readers, "understand that I merely threw out as matter for speculation, the vague impressions of my mind, for I gave nothing as the result of sufficient consideration, or as the settled conviction, or even probable conclusion at which I had arrived."

In Faraday's ether-free universe, space is suffused with electric, magnetic, and gravitational lines of force. These tensioned lines can be "plucked," in a sense, like so many strings of a cosmic violin, creating electromagnetic undulations. (Faraday suspected that gravity was linked to electricity and magnetism, and would therefore behave in similar fashion.) If a center of force—say, an electrically charged particle—oscillates up and down, then its associated lines of force will oscillate up and down as well, not instantaneously as rigid units, but with the disturbances rippling outward like waves along the lines of force. "The propagation of light," Faraday suggested, "and therefore probably of all radiant action, occupies *time;* and, that a vibration of the line of force should account for the phaenomena of radiation, it is necessary that such a vibration should occupy time also."

Neither Faraday nor anyone else in 1846 could deduce from first principles how fast such a vibration would move through space. Yet the answer is implicit in his model: If, as Faraday says, light is a vibration of lines of force, then such a vibration must move at precisely the speed of light, which had been measured at about

190,000 miles per second. But Faraday could not even verify that ray-vibrations exist, much less clock their movement.

In closing his ray-vibrations letter, the experimentalist Faraday scolds himself for having voiced such foggy ideas to the public: "I think it is likely that I have made many mistakes in the preceding pages, for even to myself, my ideas on this point appear only as the shadow of a speculation, or as one of those impressions on the mind which are allowable for a time as guides to thought and research. He who labours in experimental inquiries knows how numerous these are, and how often their apparent fitness and beauty vanish before the progress and development of real natural truth."

Coming on the heels of his revisionist ideas about the atom, Faraday's ray-vibrations paper only confirmed what the scientific community had long suspected. Faraday, the great experimentalist, was ill-equipped to tackle theoretical issues. He was trying to subvert some of the most cherished scientific paradigms, but in an arcane dialect of images and analogies that sounded discordant to mathematically trained ears.

Even Tyndall found Faraday's speculative language frustratingly vague: "It is amusing to see how many write to Faraday asking him what the lines of force are. He bewilders even men of eminence . . . I heard [French physicist Jean-Baptiste] Biot once say that he could not understand Faraday, and if you look for exact knowledge in his theories you will be disappointed—flashes of wonderful insight you meet here and there, but he has no exact knowledge himself, and in conversation with him he readily confesses this." But, in a somewhat condescending assessment, Tyndall sees a bright side to Faraday's obtuse style: "Faraday's achievements are due to his immense earnestness and great love for his subject and this very mistiness which serves to obscure the verity of matters may have its compensations by rendering the subject attractive and thus wooing a man to work at it with more fervour."

Astronomer Royal George Biddell Airy published a thoughtful

rebuttal to Faraday's ray-vibrations paper, politely forewarning Faraday of his dissent. Another defender of the status quo was not so collegial; he suggested in *The Athenaeum* that Faraday brush up on his mathematics and leave theoretical physics to the properly trained.

Faraday was unswayed by the criticisms. He had crossed a Rubicon in speculative science and had no intention of retreating. He sought to "shake men's minds from their habitual trust" in long-held but (in his opinion) specious views. "[I]t is better to be aware, or even to suspect, we are wrong, than to be unconsciously or easily led to accept an error as right." The mutability of science had been a constant for him since the beginning, inspired no doubt by the venerated guide of his youth, Isaac Watts: "Do not think learning in general is arrived at its perfection," Watts declares in *Improvement of the Mind,* "or that the knowledge of any particular subject in any science cannot be improved, merely because it has lain five hundred or a thousand years without improvement. The present age, by the blessing of God on the ingenuity and diligence of men, has brought to light such truths in natural philosophy, and such discoveries in the heavens and the earth as seemed beyond the reach of man."

As long as Faraday rendered the lines of force hypothetical—as a scheme to visualize forces, not as actual agents of force or conveyors of light—the action-at-a-distance and luminiferous ether constituencies remained largely at bay. But in his ray-vibrations paper, the lines were unmistakably real and vital to a comprehensive explanation of natural phenomena. In these *physical* lines of force, Faraday was proposing a wholly new vision of nature, which came to be known as field theory. He was shifting the focus from material entities—particles, magnets, electric circuits, planets—to the regions around them. It was here in the surrounding space that the various tugs and nudges of force are observed to act. It was here, he felt, that theorists should direct their attention.

Faraday's field of force lines constitutes a preexisting *potential* for force to occur. That potential for action is the hallmark of a field. The object at its heart—once considered to be the wellspring of force—is now inert, or at least conceptually immaterial. The surrounding field, whether envisioned concretely as lines of force or abstractly as mathematical equations, encompasses the ever-present and ever-ready power to exert its influence on any object that strays within it.

In the decade after his ray-vibrations paper, Faraday tried to secure a beachhead for field theory, as embodied in his physical lines of force. He conducted further studies of magnetic action in metals, gases, crystals, and the earth's atmosphere, in the hope of leveraging his speculative work with experimental findings. For several months, he dropped wire coils to see whether a change in gravitational force, like a change in magnetism, might lead to the induction of electricity. (It did not.) He established the quantitative relationship between the concentration of magnetic lines of force and the electrical voltage induced in a coil, now a staple of freshman physics known as Faraday's law. His reports to the Royal Institution and the Royal Society alternate between basic laboratory investigations and far-reaching speculation.

In an 1853 paper, "Observations on the Magnetic Force," Faraday utters a virtual battle cry for his case, quoting the illustrious Isaac Newton himself: "That gravity [Newton wrote to mathematician Richard Bentley] should be innate, inherent and essential to matter, so that one body may act upon another at a distance through a *vacuum,* without the mediation of anything else, by and through which their action and force may be conveyed from one to another, is, to me so great an absurdity that I believe no man who has in philosophical matters a competent faculty of thinking, can ever fall into it." (It was not Newton but his many successors who elevated action-at-a-distance to physical reality. They saw no alternative.)

Two years later, in a speculation titled "On Some Points of

Magnetic Philosophy," Faraday again wraps himself in Newton's exalted mantle. And he hints at the opprobrium his controversial stands have brought him: "At present we are accustomed to admit action at sensible distances, as of one magnet upon another, or of the sun upon the earth, as if such admission were itself a perfect answer to any inquiry into the nature of the physical means which cause distant bodies to affect each other; and the man who hesitates to admit the sufficiency of the answer, or of the assumption on which it rests, and asks for a more satisfactory account, runs the risk of appearing ridiculous or ignorant before the world of science."

Then, after reciting a litany of laboratory results, Faraday abruptly shifts gear and writes, "It is, probably, of great importance that our thoughts should be stirred up at this time to a reconsideration of the general nature of physical force, and especially to those forms of it which are concerned in actions at a distance." (The sighs of his fellow scientists are practically audible: *Here he goes again.*) Faraday considers a straightforward example: the gravitational pull of the sun on the earth. Which is more likely, he asks—that the earth's mere presence elicits gravity in the sun; or that the sun's attractive power would exist in space even if the earth did not? "I think it will be exceedingly difficult to conceive that the sudden presence of the earth, 95 millions of miles from the sun, and having no previous connexion caused by the mere circumstance of juxtaposition, should be able to raise up in the sun a power having no previous existence. As respects gravity, the earth must be considered as inert, previously, as the sun; and can have no more inducing or affecting power over the sun than the sun over it: both are assumed to be *without* power in the beginning of the case; how then can that power arise by their mere approximation or coexistence? That a body without force should raise up force in a body at a distance from it, is too hard to imagine . . ."

He goes on to argue at length against the creation of force from nothing, then poses his alternative—the gravitational field: ". . . the power is always existing around the sun and through infinite space, whether secondary bodies be there to be acted upon by gravitation or not; and not only around the Sun, but around every particle of matter which has existence." In closing, Faraday points out that all of his remarks about the utility of the field concept in gravity apply equally well to observed electrical and magnetic phenomena.

Despite repeated attempts, Faraday made little headway convincing scientists of the viability of his speculations. George Biddell Airy spoke for the scientific community when he said, "I can hardly imagine anyone who practically and numerically knows this agreement [between experiments and accepted mathematical laws], to hesitate an instant in the choice between this simple and precise action, on the one hand, and anything so vague and varying as lines of force, on the other hand." Lines of force, ray-vibrations, fields—all sounded so strange and contrived compared to more comfortable notions that held sway. "How few understand the physical lines of force!" Faraday lamented to his niece, Margery Ann Reid, in 1855. "They will not see them, yet all researches on the subject tend to confirm the views I put forth many years since."

Faraday's frustration was understandable. "The lines of force . . . stood before his intellectual eye," German physicist Heinrich Hertz remarked, ". . . as tensions, swirls, currents, whatever they might be—that he himself was unable to state—but they were there, acting upon each other, pushing and pulling bodies about, spreading themselves about and carrying the action from point to point." Faraday needed a translator. Someone who could pick up where William Thomson had left off in the 1840s. Someone who could turn his complex ideas into the hard rubric of equations and, with mathematical authority, speak out on his behalf.

By the mid-1850s, having run hard up against the limits of his own intellect, Faraday had no choice but to wait for someone else to prove—or disprove—his fantastic speculations. Until then, he could console himself with a passage he had once jotted in his diary—the knowing passage of one who had stood at the frontier of discovery and thrilled to the possibility of what lay beyond: "Nothing is too wonderful to be true, if it be consistent with the laws of nature . . ."

12

A PARTAKER OF INFINITY

Mathematics is the language with which God has written the universe.

—GALILEO

In 1841, no one at the Edinburgh Academy thought that the new boy with the square-toed shoes, blousy tunic, and frilled collar would amount to anything, much less become one of the greatest scientists of all time. Young James Clerk Maxwell—Dafty, to his sniggering classmates—was an outsider from the start, distanced as much by his rural brogue and cockeyed sense of fashion (his father's idea) as by his quicksilver train of thought. He had come into the world during the summer of 1831, almost coincident with Michael Faraday's discovery of electromagnetic induction. In London, the gifted experimentalist Faraday, having augmented his senses with implements of science, was exploring a universe of possibilities. Hundreds of miles to the north, the infant Maxwell was exploring the same universe with grasping fists and unprejudiced eyes. Man and child were as yet unaware that their paths would converge thirty years later.

Even as a child, roaming the family estate at Glenlair, on the

River Urr in southern Scotland, Maxwell sought to understand the workings of things, both natural and manmade. "What's the go o' that?" he would ask his scientifically minded father, a lawyer by profession, and if not satisfied with the explanation, would press, "But what is the *particular* go of it?" When "Jamsie" was just three, his mother Frances described his antics to her sister: "He is a very happy man, and has improved much since the weather got moderate; he has great work with doors, locks, keys, etc., and 'Show me how it doos' is never out of his mouth. He also investigates the hidden courses of streams and bell-wires . . . and he drags papa all over to show him the holes where the wires go through." By age eight, Maxwell could rattle off all 176 verses of the 119th Psalm. And like the young Humphry Davy decades earlier, he scoured the rolling countryside near his home for specimens of rocks, plants, and creatures.

Now, at ten, having spurned a private tutor, the hapless Maxwell was flung into a crucible of adolescent wretchedness at the Edinburgh Academy. Among the many torments inflicted by his classmates, in addition to tearing his clothes, was the "game" in which he was forced to jump "frog-like over a handkerchief, under the stimulus of the lash, till his exhaustion was complete." One of Maxwell's few school friends recalled that the much-put-upon Dafty "was at first regarded as shy and rather dull. He made no friendships, and he spent his occasional holidays in reading old ballads, drawing curious diagrams, and making rude mechanical models. This absorption in such pursuits, totally unintelligible to his schoolfellows (who were then quite innocent of mathematics), of course procured him a not very complimentary nickname . . ." Maxwell himself reflected laconically, "They never understood me, but I understood them."

The curriculum at the Edinburgh Academy included Latin, Greek, mathematics, and a smattering of Newtonian science. There was also a complete course in English literature and manners to "groom" the rude young Scots for entry to Cambridge or Oxford. During his first four years, Maxwell was indifferent to his studies. Then, inspired by a geometry course, "he surprised his companions," according to classmate P. G. Tait, "by suddenly becoming one of the most brilliant among them . . ." By the time he graduated in 1847, Maxwell had risen to the top of his class, garnered the Academy's awards in mathematics, English, and English verse, and even published a paper on the geometry of ovals. (The paper was read to Edinburgh's Royal Academy by a local physics professor, since Maxwell was only fourteen at the time.)

Several of Maxwell's classmates headed to Cambridge or Oxford after just one year at a Scottish university. Maxwell chose to spend a full three-year term at the University of Edinburgh. He wanted to be close to his father, who visited Edinburgh on the slightest pretext. (The two of them had attended scientific meetings together since James was twelve.) But Maxwell also needed to choose between the security of a law career like his father's or the uncertainty of the scientific path. British universities typically had only one science professorship and that opened up only upon the death of the incumbent. Nor was scientific ability the sole criterion for appointment; social connections mattered. In the end, Maxwell took the plunge. In 1850, he entered Cambridge, intent on becoming a scientist.

As Maxwell grew to adulthood, the unorthodox bent he exhibited as a youngster crystallized into full-fledged, but endearing, eccentricity. "While extremely neat in person," said a friend at Edinburgh, "he had a rooted objection to the vanities of starch and gloves. He had a pious horror of destroying anything—even a scrap of writing paper. He preferred travelling by third class in railway journeys, saying he liked a hard seat. When at table he often seemed abstracted from what was going on, being absorbed in observing the effects of refracted light in the finger-glasses, or in trying some experiment with his eyes—seeing round a corner, making invisible stereoscopes, and the like." Maxwell's Aunt Jane chided him for puzzling over mathematical propositions at the dinner table, crying, "Jamsie, you're in a prop." One of Maxwell's Cambridge professors remarked that the young Scotsman was "unquestionably the most extraordinary man he [had] met with in the whole range of his experience," in that it seemed "impossible for Maxwell to think incorrectly on physical subjects."

During holidays from school, Maxwell was industrious to a fault, as in this 1848 report to former Edinburgh Academy schoolmate Lewis Campbell: "When I waken I do so either at 5.45 or 9.15, but I now prefer the early hour, as I take the most of my violent exercise at that time, and thus am *saddened down,* so that I can do as much still work afterwards as is requisite, whereas if I was to sit still in the morning I would be yawning all day. So I get up and see what kind of day it is, and what field works are to be done; then I catch the pony and bring up the water barrel . . . Then I take the dogs out, and then look round the garden for fruit and seeds, and paddle about till breakfast-time. After that take up Cicero and see if I can understand him. If so, I read till I stick; if not, I set to [Xenophon] or [Herodotus]. Then I do props . . ."

Maxwell also immersed himself in home-brewed experiments, reminiscent of the youthful Faraday's mantelpiece investigations. He described his Glenlair "laboratory" to his friend Campbell: "I have set up shop now above the wash-house at the gate, in a garret. I have an old door set on two barrels, and two chairs, of which one is safe, and a skylight above, which will slide up and down. On the door (or table) there is a lot of bowls, jugs, plates, jam pigs [jars] etc., containing water, salt, soda, sulphuric acid, blue vitriol, plumbago ore; also broken glass, iron, and copper wire, copper and zinc plate, bees' wax, clay, rosin, charcoal, a lens, a Smee's Galvanic apparatus, and a countless variety of little beetles, spiders, and wood lice, which fall into different liquids and poison themselves . . . First, I thought a beetle was a good conductor, so I

James Clerk Maxwell in the 1850s, about the time he first wrote to Faraday.

embedded one in wax (not at all cruel, because I slew him in boiling water in which he never kicked), leaving his back out; but he would not do . . . I am reading Herodotus, Euterpe, . . . diff[erential] and Int[egral] Calc[ulus], Poisson, Hamilton's dissertations, etc. Off, then I take back to experiments, history of what you may call it, make up leeway in the newspapers, read Herodotus, and draw figures of the curves above. Oh deary, 11 P.M.!" To Cambridge, Maxwell brought not only the wealth of his undergraduate education at Edinburgh, but also, from his wash-house laboratory, his stockpile of materials for experimentation–"matter in the wrong place," sniffed one acquaintance; or in the words of his father, "Jamsie's dirt."

Compared to his Edinburgh experience, Cambridge was a pressure cooker. The curriculum focused on rote memorization of facts and established theories. Tutors guided students through solutions of textbook problems in preparation for the grueling, weeklong Tripos examination in the final year. The Scottish model, by contrast, was holistic, with strong core training in philosophy. It was from Edinburgh that Maxwell learned to identify and suppress his preconceptions; from here, too, stemmed his distrust of scientific dogma. As he told his friend Lewis Campbell, "Objectivity alone is favourable to the free circulation of the soul."

At Cambridge, Maxwell was nothing if not disciplined. Even during frigid months, he swam in the river, belly-flopping into the water to "stimulate the circulation." There was also the short-lived sleep-cycle experiment, which had him running the corridors of his dormitory at 2 A.M.–dodging boots and hair brushes hurled at him by annoyed residents. So grueling was his self-imposed study regimen that he collapsed from exhaustion–or "brain fever," in Victorian symptomatology–during the summer of 1853. He recovered and, the following winter, snagged second prize in the university's marathon Tripos mathematics competition, plus the Smith's Prize for overall excellence in mathematics. (To ward off the cold, he took the examinations with a carpet wrapped around his legs–like the square-toed shoes, his father's idea.)

Appointed a Fellow at Cambridge in 1854, Maxwell investigated a variety of real-world phenomena: the properties of fisheye lenses, the swooping motion of a descending piece of paper, and the ability of falling cats to right themselves (which triggered the false rumor that Maxwell threw cats out windows). Not the most consequential of problems, for sure, but ones that gave Maxwell a chance to exercise his mathematical muscle and his "contrarian" approach to science. The latter he declared to Lewis Campbell in a roiling olio of religious-agricultural metaphor: "Now, my great plan, which was conceived of old, and quickens and kicks periodically, and is continually making itself more obtrusive, is a plan of Search and Recovery, or Revision and Correction, or Inquisition and Execution, etc. The Rule of the Plan is to let nothing be wilfully left unexamined. Nothing is to be *holy ground* consecrated to Stationary Faith, whether positive or negative. All fallow land is to be ploughed up, and a regular system of rotation followed . . . Never hide anything, be it weed or no, nor seem to wish it hidden. So shall all men passing by pluck up weeds and brandish them in your face, or at least display them for your inspection . . . Again I assert the Right to Trespass on any plot of Holy Ground which any man has set apart . . . to the power of Darkness. Such places much be exorcised and desecrated till they become fruitful fields." Sentiments Michael Faraday would have seconded—although in less strident prose.

While Maxwell, like Faraday, barred religious doctrine from his scientific pursuits, there was a strongly spiritual element to his investigations of nature. In a private note from the mid-1850s, Maxwell reflected, "Happy is the man who can recognize in the work of To-day a connected portion of the work of life, and an embodiment of the work of Eternity. The foundations of his confidence are unchangeable, for he has been made a partaker of Infinity. He strenuously works out his daily enterprises, because the present is given him for a possession. Thus ought Man to be an impersonation of the divine process of nature, and to show forth the union of the infinite with the finite, not slighting his temporal existence, remembering that in it only is individual action possible, nor yet shutting out from his view that which is eternal, knowing that Time is a mystery which man cannot endure to contemplate until eternal Truth enlighten it."

Maxwell's research aspirations were to send him far and wide across the landscape of nineteenth-century physics. Of his influence on science, one colleague wrote, "There is scarcely a single topic that he touched upon which he did not change almost beyond recognition." During his career, Maxwell would derive the statistical laws governing the random movements of molecules in a gas; conduct pioneering studies of color vision and photography; and prove that Saturn's rings, once believed to be solid, consist of myriad particles (a feat Astronomer Royal George Biddell Airy would hail as one of the "most remarkable applications of Mathematics to physics that I have ever seen"). But the branch of science in which Maxwell would leave the biggest imprint was electromagnetism.

In February 1854, Maxwell wrote to William Thomson, who had first "mathematized" Faraday's lines of force, and asked for a reading list of great works on electricity and magnetism. Maxwell sought a path toward the observed phenomena untrammeled by doctrinaire thinking or mathematical abstraction. He wished to avoid what he termed "old traditions about forces acting at a distance" and instead tackle the subject without prejudice. Although Thomson's reply is lost, there is no doubt about his prime recommendation, for soon Maxwell was immersed in Faraday's *Experimental Researches in Electricity*. It didn't take him long to realize that this was truly "a first step in right thinking."

If not quite a revelation to Maxwell, Faraday's work nonetheless resonated with his skepticism about "Stationary Faith" in explanations of natural phenomena. "The method which Faraday employed in his researches," Maxwell wrote, "consisted in a constant appeal to experiment as a means of testing the truth of his ideas, and a constant cultivation of ideas under the direct influence of experiment . . . Faraday . . . shows us his unsuccessful as well as his successful experiments, and his crude ideas as well as his developed ones, and the reader, however inferior to him in inductive power, feels sympathy even more than admiration, and is tempted to believe that, if he had the opportunity, he too would be a discoverer. Every student therefore should . . . study Faraday for the cultivation of a scientific spirit, by means of the action and reaction which will take place between newly discovered facts as introduced to him by Faraday and the nascent ideas in his own mind."

Maxwell, the consummate mathematician, nonetheless understood the power of mathematics to mislead when not anchored in experiment or observation. In Faraday's *Researches,* he encountered science in its purest form, "untainted" by mathematical manipulation. *Here,* he decided, would be the entry point for his own investigations into electricity and magnetism. In a later reflection, Maxwell sounds almost relieved that Faraday had stuck to his particular brand of investigation, thereby blazing a trail that Maxwell himself could follow: "It was perhaps for the advantage of science that Faraday, though thoroughly conscious of the fundamental forms of space, time, and force, was not a professed mathematician. He was not tempted to enter into the many interesting researches in pure mathematics which his discoveries would have suggested if they had been exhibited in a mathematical form, and he did not feel called upon either to force his results into a shape acceptable to the mathematical taste of the time, or to express them in a form which mathematicians might attack. He was thus left at leisure to do his proper work, to coordinate his ideas with his facts, and to express them in natural, untechnical language."

Maxwell found Faraday's force-field concept a promising alternative to the blatantly mystical action-at-a-distance mechanism, especially in light of William Thomson's initial foray into the mathematics of lines of force. He decided to pick up where Thomson had left off. By the end of 1855, he had readied a paper for the Cambridge Philosophical Society in which he envisioned Faraday's lines of force as flexible, fluid-filled tubes. "I do not think that it contains even the shadow of a true physical theory," he readily admitted, viewing his model as but "a temporary instrument of research." In adopting such a physical analog, Maxwell used the laws of a well-understood science—in this case, the known properties of stationary and flowing fluids—to illuminate those of a science yet in the shadows—electromagnetism. Like Thomson before him (who had modeled lines of force as conductors of heat), Maxwell drew Faraday's vision in symbolic terms, suspending judgment as to its completeness or correctness. His particular mathematical formulation permitted him to predict, not just the broad electromagnetic effects of entire loops and coils, but even small-scale influences near tiny segments of an electric circuit. Perhaps Maxwell intuited that his technique might throw open the door to some of the most profound scientific insights of the age. During a carriage ride along the River Urr, near his Glenlair estate, an animated Maxwell described to Lewis Campbell the melding of his own vision of the lines of force with Faraday's. Campbell later remarked, "It was like listening to a fairy-tale."

13

THE LIGHT UNSEEN

*Round about the accredited and orderly facts of every science
there ever floats a sort of dust-cloud of exceptional
observations, of occurrences minute and irregular and seldom
met with, which it always proves more easy to ignore than to
attend to . . . Anyone will renovate his science who will
steadily look after the irregular phenomena, and when
science is renewed, its new formulas often have more of the
voice of the exceptions in them than of what were
supposed to be the rules.*

—WILLIAM JAMES

In early 1857, Michael Faraday drew from his mail a research paper from one *J. Clerk Maxwell, B.A., Fellow of Trinity College, Cambridge.* "On Faraday's Lines of Force" was heavy going for sixty-six-year-old Faraday, suffering still from mental fatigue and memory lapses. Yet it must have been a thrill to see his beleaguered lines of force cloaked in the armor of mathematical symbolism, barreling securely down the gauntlet of experiment and observation. "I was at first almost frightened, when I saw such mathematical force made to bear upon the subject," mused Faraday, "and then wondered to see that the subject stood it so well." Here, in this paper, was the voice he had been seeking. Indeed, his own frustration with the scientific establishment was echoed in the paper's very first sentence: "The present state of electrical science seems peculiarly unfavorable to speculation."

Unlike most of the scientific community, who considered Faraday's ideas "to be of an indefinite and unmathematical character," Maxwell divined in them the germ of mathematical thinking. To him, the lines of force were but geometry, thus capable of being translated into the language of equations. "By the method which I adopt," Maxwell writes, "I hope to render it evident that I am not attempting to establish any physical theory of a science in which I have hardly made a single experiment, and that the limit of my design is to shew how, by a strict application of the ideas and methods of Faraday, the connexion of the very different orders of phenomena which he has discovered may be clearly placed before the mathematical mind."

Faraday's mind, not being among those mathematical, was unable to fully grasp the analytic particulars that made fluid-filled tubes an effective stand-in for his lines of force. He regretted his own mathematical incapacity. "[I]f I could live my life over again," Faraday confided to a friend, "I would study mathematics; it is a great mistake not to do so, but it is too late now." Still, he was relieved to have at least one person besides William Thomson advocating his theories. Maxwell's paper seemed sound to Faraday—at least the parts he could understand. And its clear, comprehensive prose gave no cause to doubt that this young Cantabridgian knew what he was talking about. On March 25, 1857, Faraday wrote to Maxwell: "I received your paper, and thank you very much for it. I do not say I venture to thank you for what you have said about 'Lines of Force,' because I know you have done it for the interests of philosophical truth; but you must suppose it is work grateful to me, and gives me much encouragement to think on."

Energized by Maxwell's ideas, as he had once been by Thomson's, Faraday declared his intention to return to the laboratory. His goal: to experimentally verify that electrical and magnetic effects take time to move through space. A time delay was a key element in the theoretical scaffolding Faraday and now Maxwell were trying to erect. A positive result would deal a fatal blow to the action-at-a-distance theory, which posited instantaneous transmission of force between objects. Delivering that blow, however, was another matter. By what means would one sense a fleeting wave of electrical or magnetic force? Faraday ticked off the practical difficulties in his laboratory diary: "First—the quickness of the action. Second—the great distance therefore required to make the propagation such as to require sensible time—and with that great distance the rapid diminution of the magnetic action that is to be made sensible. Third—the want of instantaneous indicators of magnetic action . . . Fourth—the want of a sudden source of magnetic power . . . If, considering the reasons before given there be the least hopes of finding the time [of propagation], these hopes ought to be verified and exhausted." But by this time—1857—Faraday himself was totally exhausted. In the end, he never captured the elusive ripples of force. Nor would anyone else for another thirty years, when German physicist Heinrich Hertz created and detected them in his laboratory.

In the autumn of 1857, Faraday asked Maxwell's opinion about his latest speculation, concerning gravitational lines of force. Maxwell replied with a lengthy (nonmathematical) analysis, concluding that Faraday's concepts were fundamentally sound and that his gravitational lines could " 'weave a web across the sky,' and lead the stars in their courses . . ." Faraday wrote back to his young champion with heartfelt gratitude and an apology: "If on a former occasion I seemed to ask you what you thought of my paper, it was very wrong; for I do not think any one should be called upon for the expression of their thoughts before they are prepared, and wish to give them. I have often enough to decline giving an opinion because my mind is not ready to come to a conclusion, or does not wish to be committed to a view that may by further consideration be changed. But having received your last letter, I am exceedingly grateful to you for it . . . Your letter is to me the first intercommunication on the subject with one of your mode and habit of thinking.

It will do me much good, and I shall read and meditate it again and again . . . I hang on to your words because they are to me weighty, and . . . give me great comfort."

Faraday closes the letter with a plea: "There is one thing I would be glad to ask you. When a mathematician engaged in investigating physical actions and results has arrived at his conclusions, may they not be expressed in common language as fully, clearly, and definitely as in mathematical formulae? If so, would it not be a great boon to such as I to express them so?—translating them out of their hieroglyphics, that we also might work upon them by experiment. I think it must be so, because I have always found that you could convey to me a perfectly clear idea of your conclusions, which, though they may give me no full understanding of the steps of your process, give me the results neither above nor below the truth, and so clear in character that I can think and work from them. If this be possible, would it not be a good thing if mathematicians, working on these subjects, were to give us the results in this popular, useful, working state, as well as in that which is their own and proper with them?" Maxwell wrote to William Thomson the next day: "What a painful amount of modesty he has when he talks about things which may possibly be of a mathematical cast." For Faraday, there must have also been more than a modest amount of pain. He stood marooned while his ship of experimental science was receding across a sea of mathematical abstractions.

In 1860, having spent several years as professor of natural philosophy at Marischal College in Aberdeen, Scotland, Maxwell was recruited by King's College, London. The new position carried a hefty teaching load, which Maxwell shouldered without complaint. Still, his particular flair for research did not carry over to teaching. One student observed that "Maxwell's lectures were, as a rule, most carefully arranged and written out—practically in a form fit for printing—and he would begin reading his manuscript, but at the end of five minutes or so he would stop, remarking, 'Perhaps I might explain this,' and then he would run off after some idea which had just flashed upon his mind, thinking aloud as he covered the blackboard with figures and symbols, and generally outrunning the comprehension of the best of us."

The King's College appointment also brought Faraday and Maxwell into direct contact for the first time. Their forty-year age difference and contrary educational paths mattered little. Each honored the other for his open-mindedness and particular talents. They were both seeking to uncover the intricacies of nature's plan, one at his twilight, the other at his dawn, but allied in their shared trust in experiment and observation.

Maxwell attended the Royal Institution's Friday Evening Discourses and, in May 1861, delivered his own lecture on the theory of primary colors (at which he projected the first color photograph). At the close of one Discourse, Faraday spied Maxwell wedged in the exiting throng. "Ho, Maxwell," Faraday shouted to the young explicator of molecular motions, "cannot you get out? If any man can find a way through a crowd it should be you."

With the inspiration of Faraday now just across town, Maxwell resumed his work on field theory. The plan was the same as before: Adopt a physical analog, then apply its mathematical laws to the lines of force. Out went the fluid-filled tubes of 1855. In hummed a whirring clockwork of Lilliputian rollers and cogs—rollers to represent moving "particles" of electricity, cogs to spin out "vortices" of magnetic power. Like the previous model, there was no literal truth to Maxwell's bizarre rendering of nature as machine. Yet he mathematically demonstrated how, with suitable assumptions, this mechanical avatar generated phenomena whose quantitative behavior matched those of electromagnetism, including electrical induction; the relation between optical and electrical properties of matter; and the magnetic rotation of a light beam's polarization, observed by Faraday in 1845. But Maxwell's most astounding insight had to do with the propagation of electromagnetic effects. In order to better mimic the properties of matter, Maxwell had endowed his fictitious rollers and cogs with a measure of elasticity; they could deform and also slip slightly out of position. Thus the displacement of one cog triggered a similar displacement in its neighbors. These, in turn, drew their neighboring cogs aside, and so on. The result was a sequence of disturbances that rippled through Maxwell's fantastic universe. A mechanical wave. Or, in its real-life incarnation, an *electromagnetic* wave.

While summering at Glenlair in 1861, Maxwell derived the mathematical formula that predicted the speed at which such an electromagnetic wave would move through space. But he was unable to solve the formula until he had "plugged in" crucial experimental data regarding certain properties of matter. The necessary data lay among his papers in London. Upon his return to the city that autumn, he immediately looked up the information and churned out the answer. Electromagnetic waves, according to Maxwell's calculation, sweep through the air at a speed of 193,088 miles per second—almost precisely the speed of light, as it was then known. In essence, Maxwell had mathematically "shaken" an electrically charged particle—one of the fictitious rollers of his model—and found that it radiated electromagnetic energy. It produced *light*.

On October 19, 1861, Maxwell dashed off a note to Faraday with the good news: "I think we have now strong reason to believe, whether my theory is a fact or not, that the luminiferous and the electromagnetic medium are one." In other words, light is indeed an electromagnetic undulation—a "ray-vibration," as Faraday had called it in 1846. To the world, Maxwell announced, "The electromagnetic theory of light, as proposed by [Faraday], is the same in substance as that which I have begun to develop in this paper, except that in 1846 there were no data to calculate the velocity of propagation."

Maxwell released his results in an 1862 paper titled "On the Physical Lines of Force"—an homage to Faraday's 1852 paper of similar name. Yet even as this paper was in press, Maxwell set about to dismantle the rollers and cogs. His 1864 paper, "A Dynamical Theory of the Electromagnetic Field," presents his revolutionary synthesis of electrical, magnetic, and optical phenomena. Gone entirely is the visual anchor of a physical analog. In its place is a purely mathematical construction—a set of equations that completely define properties of electromagnetic fields arising from electric charges, currents, and magnets. Here was Michael Faraday's field—his lines of force—reduced to its abstract essentials. Reduced, in the cruelest of ironies, to a form Faraday could never understand.

An electromagnetic field, in Maxwell's conception, is some fundamental alteration of space wrought by embedded electric and magnetic sources. Maxwell's equations do not reveal what an electromagnetic field *is*, just how to compute its mathematical properties and how these properties give rise to observable phenomena. Inspired by Faraday's geometrical musings, Maxwell created an electromagnetic universe that cannot be effectively reduced to mental images. All self-imagined analogs to visualize the field are in some way deficient. Yet the mathematical rendering of the field is complete and accurate. Maxwell likened the situation to that of a bell ringer who tugs ropes that dangle through holes in the ceiling of the belfry; the bells themselves and their actuating mechanism remain a mystery. Maxwell's contemporary, Heinrich Hertz, put it more bluntly: "Maxwell's theory is Maxwell's equations." Or in the words of Nobel prize–winning physicist Richard Feynman, nearly a century later, "Today, we understand better that what counts are the equations themselves and not the model used to get them. We may only question whether the equations are true or false. This is answered by doing experiments, and untold numbers of experiments have confirmed Maxwell's equations. If we take away the scaffolding he used to build it, we find that Maxwell's beautiful edifice stands on its own. He brought together all of the laws of electricity and magnetism and made one complete and beautiful theory."

James Clerk Maxwell.

In the ensuing decades, Maxwell's field concept would be expanded and formalized by others, including Einstein, who derived the analogous field equations for gravity. (A variation of these large-scale field theories—called quantum field theory—would later be created to account for observed behaviors on the atomic-size scale.) And Maxwell's prediction of invisible forms of electromagnetic energy—longer- or shorter-wavelength "cousins" of visible light—would be confirmed when Hertz produced radio waves from electrical sparks in 1888. As Hertz was to point out in his paper outlining this discovery, "It is clear that the experiments amount to so many reasons in favor of that theory of electromagnetic phenomena which was first developed by Maxwell from Faraday's views."

In his 1873 masterwork, *A Treatise on Electricity and Magnetism,* Maxwell acknowledges his debt to Faraday: "I would recommend

Faraday holding a bar magnet.

the student, after he has learned, experimentally if possible, what are the phenomena to be observed, to read carefully Faraday's *Experimental Researches in Electricity*. He will there find a strictly contemporary historical account of some of the greatest electrical discoveries and investigations, carried on in an order and succession which could hardly have been improved if the results had been known from the first, and expressed in the language of a man who devoted much of his attention to the methods of accurately describing scientific operations and their results . . . If by anything I have here written I may assist any student in understanding Faraday's modes of thought and expression I shall regard it as the accomplishment of one of my principal aims—to communicate to others the same delight which I have found myself in reading Faraday's *Researches*."

14

THE SIMPLEST EARTHLY PLACE

Virtue is like a rich stone—best plain set.

—FRANCIS BACON, *OF BEAUTY*

The glittering Thames that Michael Faraday once knew had, by the 1850s, become an open-air sewer. Untreated waste poured into the river in increasing amounts as the population of London swelled. The once ubiquitous salmon were long gone. An 1853 cholera epidemic was traced to the main public water supply: the Thames.

On a sun-charged July afternoon in 1855, Faraday stared in disgust from the deck of a steamboat as it plied the Thames between London and Hungerford Bridges. From ship to shore, upstream to downstream, all was sloshing brown fluid. The low tide so concentrated the stench that it crept over the railing like a malodorous spirit. That Faraday had just returned from the unspoiled waters and pristine air of the country made the environmental havoc before him even more of an affront to his senses. Did nobody care? Could not the same authority who drains putrescent ponds in neighborhoods do something to keep London's great river clean? Faraday formulated a plan. He would write a letter—a public letter to *The Times,* expressing his outrage. It was his duty, he felt, both as a citizen and as a professed advocate of the societal benefits of science, to record what he saw. He would use the accumulated weight of his own reputation to effect change.

Faraday pulled some white cards from his pocket and ripped them into pieces. While the boat idled at St. Paul's Wharf, he dropped a few of the little paper rectangles into the Thames. They disappeared into the murkiness before they had sunk an inch. In fact, the bottoms of the pieces that had entered the water edgewise were invisible even as their white tops protruded from the surface. The water was indeed opaque. Faraday repeated his ad hoc experiment at Blackfriar's Bridge, Temple Wharf, Southwark Bridge, and Hungerford Bridge. The result was the same. The Thames was suffused everywhere with waste, some of which—Faraday duly noted— "rolled up in clouds so dense that they were visible at the surface . . ."

At Hungerford Bridge, Faraday stepped off the boat and fled the pier for the "sweeter atmosphere" of the streets. Arriving home, he immediately set to work on his letter. There would be nothing figurative in its descriptions, he decided, no exaggeration of what he had seen—just the simple truth. The letter appeared in *The Times* two days later, on July 9, 1855. "If we neglect this subject," Faraday warned, "we cannot expect to do so with impunity; nor ought we to be surprised if ere many years are over, a hot season give us sad proof of the folly of our carelessness." Faraday's warning proved prophetic: Just three years later, during a particularly hot summer, a septic miasma overran the riverfront. Scented sheets were hung in the windows of Parliament and tons of "purifying" agents were dumped into the Thames, to no avail. A number of sewage-treatment schemes were implemented in the decades after the "Great Stink," as the 1858 event came to be known, but not until the 1970s was the Thames restored to its former purity.

A few weeks after Faraday's letter appeared, *Punch* featured a full-page cartoon of him leaning over the side rail of a boat, nostrils pinched shut, offering his business card to a muck-draped Father Thames. "And we hope the Dirty Fellow will consult the learned Professor," proclaims the caption. To the general public, Faraday had become the very embodiment of science—*their* representative to the natural world around them.

The Thames issue was just the latest way in which Faraday waded more overtly into public affairs during the sunset years of his career. Now in his sixties, he was increasingly hobbled by the cognitive misfires of his brain. He embraced the inevitability of his new role as elder statesman of science. Faraday desired more than anything to be useful, if not in the laboratory or in the realm of theoretical physics, then as environmental watchdog, educator, or promoter of the societal benefits of science.

Nothing raised Faraday's hackles more than the public's easy acceptance of pseudoscience and spiritualism, which were rampant in mid–nineteenth-century England: Tables spun of their own accord; spirits visited from the afterlife; ordinary citizens claimed magical powers of perception. Some attributed the strange occurrences to electricity or magnetism, others to the intervention of previously unknown forces, still others to "diabolical or supernatural agency." And no one was subjecting the wild claims to scientific scrutiny. Yes, Faraday acknowledged, there would always be charlatans, preying on the gullible. Yet the victims themselves are culpable for their uncritical belief in the frauds being perpetrated. In a public statement with a painfully modern resonance, Faraday asked, "Shall we educate ourselves in what is known, and then casting all away we have acquired, turn to our ignorance to guide us among the unknown?"

While pseudoscience grated against Faraday's rational view of nature—that all natural phenomena and processes operate under *law*—spiritualism offended his religious sensibilities. He claimed an absolute distinction between religious and secular belief, saying in an

1854 lecture that he had "never seen anything incompatible between those things of man which can be known by the spirit of man which is within him, and those higher things concerning his future, which he cannot know by that spirit." In Faraday's opinion, God would simply not permit humans to "peek" into the afterlife, nor would elements from that unknowable realm seep into day-to-day mortal existence. The living should ask no proof of the beneficence of God or of the promise of an afterlife. Faith—pure, heartfelt faith—Faraday believed, is all the assurance one should need regarding the world beyond. In a nod to Revelation, he saw in the practitioners of the spiritual arts the influence of "unclean spirits . . . waking in the hearts of men."

Among Faraday's many correspondents, John Allen, archdeacon of Salop, urged the immediate study of table turning, which he insisted "will be ranked among the most astonishing discoveries of this age." Echoing similar accounts in contemporary newspapers, Allen described the spontaneous movement of a small, circular table at the Prussian ambassador's house in London: Eight guests had rested their hands in a circular chain on the tabletop; after some twenty minutes, the table began to move, according to Allen, its velocity "appearing to increase equably so that the experimenters were obliged to run round with it." And, Allen added, every time the chain of hands was broken, the movement miraculously ceased.

Within the week, a second missive arrived for Faraday, this time from William Edward Hickson, editor of the *Westminster Review,* in Fairseat, Wrotham, Kent. "Are we not on the eve of some new discovery in Dynamics?" Hickson wrote, then related his own table-turning experiences, including this tidbit about one paranormally challenged household: "Returning home I called my friend Fowler's the architect, whose family had been busy all the evening trying similar experiments. They had fail'd to move a table but were succeeding with a hat when I entered the room." Desperate to gain Faraday's attention, Hickson even offered the services of the sixty "inmates" of Miss Grant's school who could evidently "produce all the phenomena any evening with as much certainty as making the wheels of a clock turn by winding it up." A third table turner—"not a showman, but simply a bankers' clerk"—assured Faraday "in plain English—that no money or present of any kind would be accepted from any one however pleased and satisfied with the demonstration he might be."

Privately, Faraday was frank in his opinions of the table turners. To his niece, Caroline Deacon, he wrote, "[T]he world is running mad after the strangest imaginations that can enter the human mind. I have been shocked at the flood of impious & irrational matter which has rolled before me in one form or another . . ." And this, to his friend Schoenbein: "I declare that taking the average of many minds that have recently come before me (and apart from that spirit which God has placed in each) and accepting for a moment that average as a standard, I should far prefer the obedience affections & instinct of a dog before it. Do not whisper this however to others."

The gathered offenses of pseudoscientists, mediums, clairvoyants, *and* the credulous public were enough to induce the publicity-shy Faraday to come out of his corner swinging. "I do not object to table-moving, for itself," Faraday announced in an 1854 lecture, "for being once stated, it becomes a fit, though a very unpromising subject for experiment; but I am opposed to the unwillingness of its advocates to investigate; their boldness to assert; the credulity of the lookers-on; their desire that the reserved and cautious objector should be in error; and I wish, by calling attention to these things, to make the general want of mental discipline and education manifest."

No one was too young to join the battle against pseudoscience. With Churchillian urgency, Faraday encouraged the children at his

1853 Christmas lecture to resist the swelling tide of ignorance and superstition:

> Study science with earnestness—search into nature—elicit the truth—reason on it, and reject all which will not stand the closest investigation. Keep your imagination within bounds, taking heed lest it run away with your judgment. Above all, let me warn you young ones of the danger of being led away by the superstitions which at this day of boasted progress are a disgrace to the age, and which afford astonishing proofs of the vast floods of ignorance overwhelming and desolating the highest places.

> Educated man, missing the glorious gift of reason which raises him above the brute, actually lowers himself well below the creatures endowed with only instinct; inasmuch as he casts aside the natural sense, which might guide him, and in his credulous folly pretends to dissever and investigate phenomena which reason would not for a moment allow, and which, in fact, are utterly absurd.

> Let my young hearers mark and remember my words. I desire that they should dwell in their memory as a protest uttered in this institution against the progress of error. Whatever be the encouragement it may receive elsewhere, may we, at any rate, in this place, raise a bulwark which shall protect the boundaries of truth, and preserve them uninjured during the rapid encroachments of gross ignorance under the mask of scientific knowledge.

To society at large, Faraday sought to prevail by "turning the tables upon the table turners," as he told Schoenbein with undisguised relish. Any movement of the table, he believed, must be due to voluntary or involuntary pushes from the table turners themselves. Thinking experimentally, he constructed a sensitive tabletop indicator lever that magnified surreptitious forces exerted by the hands. "If the hands involuntarily moved to the left *without* the table," he explained in an open letter to *The Times* in June 1853, "the index would go to the right; and, if neither table nor hands moved, the index would itself remain immoveable . . . [T]he most valuable effect of this test-apparatus (which was afterwards made more perfect and independent of the table) is the corrective power it possesses over the mind of the table-turner. As soon as the index is placed before the most earnest, and they perceive—as in my presence they have always done—that it tells truly whether they are pressing downwards only or obliquely, then all effects of table-turning cease, even though the parties persevere, earnestly desiring motion, till they become weary and worn out."

But it was Faraday, in the end, who grew weary and worn out. His *Times* letter had no effect on the tide of pseudoscientific and spiritualistic claptrap, and he made no more experiments or public pronouncements on the subject. Still the letters kept coming. Some he answered; some he did not. Among the last, from a Thomas S–, Esq., elicited this reply: "Sir,—I beg to thank you for your papers, but have wasted more thought and time on so-called spiritual manifestation than it has deserved. Unless the spirits are utterly contemptible, *they* will find means to draw my attention . . . If I could consult the spirits, or move them to make themselves honestly manifest, I would do it. But I cannot, and am weary of them."

Faraday closed his table-turning letter to the *Times* with an indictment of England's educational system: "I have been greatly startled by the revelation which this purely physical subject has made of the condition of the public mind . . . I think the system of education that could leave the mental condition of the public body in the state in which this subject has found it must have been greatly deficient in some very important principle." If Faraday couldn't stop the table turners and their ilk, perhaps he could buttress society against their influence through improved scientific education for all.

Throughout the first half of the nineteenth century, the standard English education consisted of the classics of literature plus "pure" mathematics, which included astronomy. The sciences—chemistry, physics, biology, geology—were typically not taught, and if they were, the lessons were brief at best. Using himself as a prime example, Faraday held that training in the sciences, especially in the scientific method, was the best way to teach a student to properly exercise critical judgment over the imagination and the senses. In the 1850s, he began to use his prestige to lobby for inclusion of the sciences in the secondary-level curriculum and to strengthen its teaching in postsecondary institutions.

Faraday outlined his philosophy of education in an 1854 lecture at the Royal Institution. He reiterated his long-held view that "education has for its first and its last step *humility*. It can commence only because of a conviction of deficiency; and if we are not disheartened under the growing revelations which it will make, that conviction will become stronger unto the end ... The first step in correction is to learn our deficiencies, and having learned them, the next step is almost complete: for no man who has discovered that his judgment is hasty, or illogical, or imperfect, would go on with the same degree of haste, or irrationality, or presumption as before." Whatever systemic reforms might be implemented, Faraday continued, there was always a place for self-education like his own. "It is necessary that a man *examine himself,* and *that* not carelessly. On the contrary, as he advances, he should become more and more strict, till he ultimately prove a sharper critic to himself than any one else can be; and he ought to intend this, for, so far as he consciously falls short of it, he acknowledges that others may have reason on their side when they criticise him." And in a touching reference to his own experience, he lamented how the self-educated are often derided by the university-educated.

In an 1858 Discourse on the electric telegraph, Faraday advocated for full recognition of science as a branch of education. "The development of the applications of physical science in modern times has become so large and so essential to the well-being of man that it may justly be used, as illustrating the true character of pure science, as a department of knowledge, and the claim it may have for consideration by Governments, Universities, and all bodies to whom is confided the fostering care and direction of learning. As a branch of learning, men are beginning to recognize the right of science to its own particular place;—for though flowing in channels utterly different in their course and end to those of literature, it conduces not less, as a means of instruction, to the discipline of the mind; whilst it ministers, more or less, to the wants, comforts, and proper pleasure, both mental and bodily, of every individual of every class in life."

Four years later, in testimony before the public school commissioners, Faraday reemphasized the need to introduce science into the precollege curriculum of every school: "That the natural knowledge that has been given to the world in such abundance during the last fifty years, I may say, should remain untouched, and that no sufficient attempt should be made to convey it to the young mind, growing up and obtaining its first views of these things, is to me a matter so strange that I find it difficult to understand."

Faraday insisted that the problem of science education was systemic, affecting citizens of all classes: "[T]hey come to me and they talk to me about things that belong to natural science; about mesmerism, table turning, flying through the air, about the laws of gravity; they come to me to ask me questions, and they insist against me, who think I know a little of these laws, that I am wrong and they are right, in a manner that shows how little the ordinary course of education can teach these minds ... They are ignorant of their ignorance at the end of all that education ... and I say again there must be something wrong in the system of education which leaves minds, the highest taught, in such a state." Faraday envisioned a scientifically literate populace, imbued from the earliest age with lessons in observation and critical judgment. But first, he testified, society must come to "honour and encourage natural knowledge"; only then would a corps of competent science teachers arise. Faraday's lofty goals came crashing down once he realized, to his dismay, that the commissioners' reform agenda was half-hearted at best and targeted only the educational bastions of the upper class. He left the proceedings feeling ill-used. He recognized the irony that he—the epitome of self-education and merit-based advancement—had been tapped to prop up the Etons of the system, while common schools languished.

The Sandemanian church continued to hold a central place in Faraday's life. He attended services and ritual feasts, enjoyed the sense of community and, with rare exception, clung to its precepts. However, the Sandemanian universe could also be a harsh environment for an independent thinker. In 1850, Faraday took issue with church elders on a matter of scriptural interpretation: Does an individual have the power to forgive a transgressor, as Faraday held, or is that power vested solely in the church? The incident followed a similar one in 1844 when Faraday and eighteen others, including his brother, sister-in-law, and father-in-law, had been excluded from the congregation for several weeks over a matter of governance. The stakes were higher this time: According to Sandemanian law, a second exclusion was permanent.

Faraday poured out his emotions in a confused, scripture-laden note to family friend and fellow Sandemanian, William Buchanan: "I am in deep distress and I write to you for my heart is full . . . bear with my anguish & do not refuse to sympathize a little with me by receiving this patiently though I be utterly unworthy. I may well fear that a deceived heart hath turned me aside for where my only comfort ought to be there is my sore grief & trouble."

Sarah's concern was no less than her husband's, for the bonds of marriage and religion tugged her in opposite directions. "I rise from a restless bed . . ." she penned to Buchanan at 3 A.M. one night, "feeling sure of your kind sympathy in my great anxiety & affliction. My beloved husbands mind has been much disturbed in the view the Church takes of not receiving an excommunicant more than once—& he has had some conversation with our Elders which so far does not seem very satisfactory."

On November 6, 1850, in what must have been a pitiful spectacle, Faraday recanted his views before the Sandemanian elders and was spared excommunication. "God has wrought with me," he informed Buchanan, "when I was against him & broken down my pride & false reasoning and has this evening shewn me what love there is in the Church to an erring brother . . ." Faraday's life was completely compartmentalized. So, too, evidently was his person. In both the internal and external worlds, he had erected a wall separating matters secular from those religious. One day he might stand tall in the arena of science or public affairs, asserting his independence before the weight of authority, yet the next, submit meekly to the reprimands of his church. The wonder is how he managed to occupy both sides of the wall simultaneously.

Meanwhile, Faraday's mental abilities continued to deteriorate. No hypochondriac, he became instead the objective witness to his own "gentle decay," as he stoically put it. The experimenter's lens that he applied to the external world, he now held up to his own mentations. In a slackening cycle, he used his diminishing faculties to assess those same faculties—a cognitive mainspring unwinding to flaccid equilibrium.

"My memory wearies me greatly in working," Faraday confessed to the Royal Institution's secretary, the Reverend John Barlow, in 1857, "for I cannot remember from day to day the conclusions I come to, and all has to be thought out many times over. To write it down gives no assistance, for what is written down is itself forgotten. It is only by very slow degrees that this state of mental muddiness can be wrought either through or under; nevertheless, I know that to work somewhat is far better than to stand still, even if nothing comes of it."

Despite his declining abilities, Faraday managed to churn through a diverse array of research projects during the 1850s and early 1860s: long-distance telegraphy; discharge of electricity through gases; properties of chemical suspensions; conservation of paintings at the National Gallery and historical artifacts at the British Museum; lighthouse illumination and maintenance; chemical warfare (he recommended against it on practical grounds); and the design of prefabricated hospital wards for the Anglo-French war against Russia. A more speculative study involved another attempt to convert gravity into electricity, a key offshoot of Faraday's ideas about the underlying equivalence of nature's forces. Dropping heavy bricks from a great height, he suggested in his notes, might electrify them to a measurable degree. After considering the Tower of Parliament, Faraday released a 280-pound lead brick repeatedly from the Shot Tower near Waterloo Bridge–to no effect.

Faraday's laboratory diary provides a window onto his creative process in the late stages of his career. Here he is, carrying on a dialogue with his internal muse, urging ideas to flow forth from the constricted spigot of his imagination: "Surely the force of gravitation and its probable relation to other forms of force may be attacked by experiment. Let us try to think of some possibilities . . . Must not be deterred by the old experiments . . . Let us encourage ourselves by a little more imagination prior to experiment . . . Let the imagination go, guarding it by judgment and principle, but holding it in and directing it by *experiment*." But Faraday's self-exhortations proved increasingly fruitless. With each year, his experimental paths narrowed and eventually disappeared into the dense undergrowth of his own confusion.

That Faraday remained *physically* robust is evident from a report he filed when he was sixty-nine regarding lighthouse inspections: "I went to Dover last Monday (the 13th instant); was caught in a snow storm between Ashford and Dover and nearly blocked up in the train; could not go to the lighthouse that night; and finding, next day, that roads on the downs were snowed up, returned to London. On Friday I again went to Dover and proceeded by a fly that night, hoping to find the roads clear of snow; they were still blocked up towards the lighthouse, but by climbing over hedges, walls, and fields, I succeeded in getting there and making the necessary inquiries and observations." Faraday retired from lighthouse work in 1865, at age seventy-four.

Faraday's social life expanded somewhat in the 1850s. He and Sarah attended the theater more frequently, as well as various institutional banquets, plus occasional soirées at the homes of some of London's most prominent families. They themselves rarely entertained, except with family or the few friends and colleagues they permitted into the inner sanctum of their apartment at the Royal Institution. John Tyndall recorded one such visit in 1853, a rare dinner gathering when Sarah was away: "At two o'clock he came down for me. He, his niece, and myself, formed the party. 'I never give dinners,' he said. 'I don't know how to give dinners, and I never dine out. But I should not like my friends to attribute this to a wrong cause. I act thus for the sake of securing time for work, and not through religious motives, as some imagine.' He said grace. I am almost ashamed to call his prayer a 'saying of grace.' In the language of Scripture, it might be described as the petition of a son, into whose heart God had sent the Spirit of His Son, and who with absolute trust asked a blessing from his father. We dined on roast beef, Yorkshire pudding, and potatoes; drank sherry, talked of research and its requirements, and of his habit of keeping himself free from the distractions of society. He was bright and joyful—boy-like, in fact, though he is now sixty-two."

The famous German physicist Hermann von Helmholtz also visited that year, and remarked: "[Faraday] is as simple, charming, and unaffected as a child; I have never seen a man with such winning ways. He was, moreover, extremely kind, and showed me all there was to see. That, indeed, was little enough, for a few wires and some old bits of wood and iron seem to serve him for the greatest discoveries."

In addition to their getaways in the country or at the coast, the Faradays rented various residences in the greater London area, so they could escape the hectic work environment and the urban smog. But once a neighborhood learned that the esteemed professor was in their midst, social invitations poured in, and the Faradays moved on.

As age overtook him, Faraday was forced to shed his most cherished professional activities. On October 11, 1861, he informed the managers of the Royal Institution that he had delivered his last Christmas lecture for the young:

It is with the deepest feeling that I address you.

I entered the Royal Institution in March 1813, nearly forty-nine years ago, and, with exception of a comparatively short period, during which I was absent on the Continent with Sir Humphry Davy, have been with you ever since.

During that time I have been most happy in your kindness, and in the fostering care which the Royal Institution has bestowed upon me. I am very thankful to you, and your predecessors for the unswerving encouragement and support which you have given me during that period. My life has been a happy one and all I desired. During its progress I have tried to make a fitting return for it to the Royal Institution and through it to Science.

But the progress of years (now amounting in number to threescore and ten) having brought forth first the period of development, and then that of maturity, have ultimately produced for me that of gentle decay. This has taken place in such a manner as to render the evening of life a blessing:—for whilst increasing physical weakness occurs, a full share of health free from pain is granted with it; and whilst memory and certain other faculties of the mind diminish, my good spirits and cheerfulness do not diminish with them.

Still I am not able to do as I have done. I am not competent to perform as I wish, the delightful duty of teaching in the Theatre of the Royal Institution, and I now ask you (in consideration for me) to accept my resignation of the *Juvenile lectures* . . . I may truly say, that such has been the pleasure of the occupation to me, that my regret must be greater than yours need or can be.

The following June, Faraday presented his final Friday Evening Discourse, on gas furnaces. His lecture notes bear the imprint of his growing infirmity: scorch marks when he evidently fumbled his papers near a flame. At the end of his notes, he sets out his retirement announcement:

Personal explanation,—years of happiness here, but time of retirement; LOSS OF MEMORY and *physical endurance of the brain.*

1. Causes—*hesitation and uncertainty* of the convictions which the speaker has to urge.
2. *Inability to draw* upon the mind for the treasures of knowledge it has previously received.
3. *Dimness,* and forgetfulness of one's former *self-standard* in respect of *right, dignity,* and *self-respect.*
4. Strong duty of *doing justice to others,* yet inability to do so.

Retire.

March 12, 1862, marked Faraday's last experimental foray, an unsuccessful attempt to demonstrate the influence of magnetism on the light-spectrum of incandescent substances. Thirty-five years later, figuring Faraday might have been onto something, Dutch physicist Pieter Zeeman took up the experiment with the more sensitive apparatus of the day. His observation of magnetically induced doubling of spectral features brought him the Nobel Prize in 1902, an honor for which Zeeman justly credited Faraday.

In 1862, Faraday and his wife took up residence in a comfortable house at Hampton Court, courtesy of Queen Victoria (at Prince Albert's urging). Although still officially on staff at the Royal Institution and, to a degree, still professionally active, Faraday felt the world closing in on him. Two years earlier, he had complained to Schoenbein: "When I try to write [to you] of science, it comes back to me in confusion; I do not remember the order of things, or even the facts themselves . . . ; and if I try to remember up, it becomes too much, the head gets giddy, and the mental view only the more confused." Now, suffering another memory crisis, he composed the letter to Schoenbein that would terminate their decades-long correspondence: "Again and again I tear up my letters, for I write nonsense. I cannot spell or write a line continuously. Whether I shall recover—this confusion—do not know. I will not write any more. My love to you." Schoenbein, at a loss for what to say, never responded.

As always, Sarah provided a safe harbor for Faraday's meandering mind. "My head is full, and my heart also," he wrote her while traveling to Glasgow in 1863, "but my recollection rapidly fails, even as regards the friends that are in the room with me. You will have to resume your old function of being a pillow to my mind, and a rest, a happy-making wife." By year's end, he confided to a friend, matter of factly, that he was "altogether a very tottering and helpless thing."

Increasingly, Faraday sat virtually immobile and silent in a chair overlooking the grounds of Hampton House—a western-facing chair so he could watch the sunset. In 1866, chemist Henry Roscoe stopped by to ask whether he might borrow a suspension of gold particles Faraday had created years earlier in his lab. "His mind was then failing," Roscoe reported, "and it was quite sad to see that he hardly understood what I was asking for. Mrs. Faraday, who stood close by, tried to recall the facts to his mind and said to him: 'Dont you remember those beautiful gold experiments that you made?' To which he replied in a feeble voice: 'Oh, yes, beautiful gold, beautiful gold,' and that was all he would say." In a lucid moment, Faraday told astronomer James South that he wished to "have a plain simple funeral, attended by none but [his] own relatives, followed by a gravestone of the most ordinary kind, in the simplest earthly place." On the afternoon of August 25, 1867, Faraday died while sitting in his chair. His relatives conveyed his coffin to Highgate Cemetery. During the private burial, several colleagues "came out from the shrubbery" to pay their respects.

EPILOGUE

We shall not cease from exploration,
And the end of all our exploring,
Will be to arrive where we started,
And know the place for the first time.

—T. S. Eliot, "Little Gidding" (from "The Four
Quartets" in *The Collected Poems of T. S. Eliot*,
1934, 1936)

There is a serenity in knowing that nature is explicable and beck-
ons generation after generation to know it better. Michael Faraday
sought to understand the natural world on behalf of us all, in the belief that the revealed knowledge would nourish the collective soul of
humanity. As Faraday recognized, we are the eyes of the universe
gazing upon itself, absorbing into our consciousness its vast and intricate puzzle. He lived to decode that puzzle.

During a career that spanned more than four decades, Faraday
laid the experimental foundations of our technological society;
made important advances in chemistry, optics, geology, and metal-
lurgy; developed prescient theories about space, force, and light;
pressed for a scientifically literate populace years before science had
been deemed worthy of common study; and manned the barri-
cades against superstition and pseudoscience. He sought no finan-
cial gain or honorifics from any of his discoveries.

At a time when mathematics was fast becoming the analytical

tool of theoretical physics, Faraday put forth revolutionary ideas in the only way he could—through the skilled use of intuition, logic, and language. The scientific establishment viewed these "ramblings" with bemusement, if not outright scorn. It was not until Maxwell translated Faraday's ideas into the hard dialect of equations that their essential correctness came to be recognized. Ultimately, Einstein would characterize Faraday's and Maxwell's electromagnetic theory as the "greatest alteration . . . in our conception of the structure of reality since the foundation of theoretical physics by Newton."

Despite the magnitude of his undertaking, Michael Faraday never abandoned his childlike sense of wonder. It was with unfettered joy that he approached his "work," for he knew that the creation of the divine hand was everywhere to be seen, to be investigated, to be understood. That joy bubbled to the surface one day in the late 1850s when physicist Julius Plücker of Bonn came to the Royal Institution to demonstrate a new vacuum tube he had developed. With Faraday at his side, Plücker applied a high voltage to the tube's metal electrodes. From within rose a greenish glow that bathed the scientists' faces in its ethereal light. Plücker then brought a magnet up to the glass—the discharge in the tube flared into a miniature aurora. Faraday, rippling with glee, danced around the table. He spoke for every creative mind, every devotee of nature's sublime spectacle, every curious spirit, whether six or sixty, when he cried: "Oh, to live always in it!"

ABBREVIATIONS

BJI & II   Jones, Henry Bence, *The Life and Letters of Faraday,* vols. I & II.

C&G   Campbell, Lewis, and William Garnett, *The Life of James Clerk Maxwell, with Selections from His Correspondence and Occasional Writings.*

Diary   Faraday, Michael, *Faraday's Diary, Being the Various Philosophical Notes of Experimental Investigation,* edited by Thomas Martin.

ERE   Faraday, Michael, *Experimental Researches in Electricity.*

NOTES

<small>PREFACE</small>

xii "conviction of deficiency": Faraday, "Observations on Mental Education," May 6, 1854; Williams (1965), p. 338.
xiii "disgrace to the age": BJ II, p. 309.

<small>CHAPTER 1: IMPROVEMENT OF THE MIND</small>

5 "searching for some Mineral or Vegitable curiosity . . .": Williams (1965), p. 11.
5 "I set off from you at a run . . .": Faraday to Benjamin Abbott, July 20, 1812; Williams (1965), p. 23.
6 "He drinks from a fount on Sunday . . .": Williams (1965), p. 6, from Tyndall's Journals at the Royal Institution.
6 "Even the lower orders of men . . .": Watts (1809), p. 13.
7 "Do not content yourselves with mere words . . .": Watts (1809), p. 119.
7 "Let not young students apply themselves . . .": Watts (1809), p. 119.
7 "that of the most illiterate": Faraday (1835), p. 81.
8 "a man who, though his conduct . . .": Williams (1965), p. 14.
9 "There is something more sprightly . . .": Watts (1809), p. 33.
11 "To you . . . is to be attributed . . .": Gladstone (1873), p. 5.
11 "Spirit of Party and bigotry . . .": Williams (1960b), p. 522.
13 "I assure you that the most wonderful . . .": Lindee (1991), p. 11.
13 "a lesson of piety and virtue": Lindee (1991), p. 21.

13 "Do not suppose that . . .": Faraday to Auguste de la Rive, October 2, 1858; Williams (1960b), p. 522.

15 "a kind of sublime *Mechanics' Institute . . .*": Foote (1952), p. 7.

16 "the power, wisdom, and goodness . . .": Forgen (1980), p. 195.

16 "mere points for employing the lever . . .": Forgen (1980), p. 194.

18 "The apparatus . . . which will, no doubt, astonish you . . .": Mertens (1998), p. 302.

19 ". . . a very disagreeable quivering and pricking"; ". . . a sensation of light in the eyes . . ."; ". . . a kind of crackling with shocks . . .": Mertens (1998), pp. 302–303.

19 "I, Sir, I my own self, cut out seven discs . . .": Faraday to Benjamin Abbott, July 12, 1812; Williams (1965), p. 22.

20 "[B]oth wires became covered in a short time . . .": Faraday to Benjamin Abbott, July 12, 1812; BJI, p. 20.

20 "[O]n separating the discs [of the battery] from each other . . .": Faraday to Benjamin Abbott, July 12, 1812; Williams (1965), p. 22.

21 "I was never able to make a fact my own . . .": Faraday to Dr. Becker, October 25, 1860; Williams (1965), p. 27.

21 "the binding of other men's thoughts . . .": Crosse (1891), p. 37.

22 "however menial"; "the letter required no answer": Faraday (1835), p. 82.

CHAPTER 2: PERCEPTIONS PERFECTLY NOVEL

24 "Upon this principle . . .": Glashow (1994), p. 197.

25 "I was wrong in publishing a new theory . . .": Hartley (1966), p. 19.

25 "persons with Consumption . . .": Cooper (2000), p. 920.

26 "produced a spasm . . .": Cooper (2000), p. 920.

26 "A thrilling extending from the chest . . ."; "Nothing exists but thoughts!": from Davy, *Researches . . .* , cited in Knight (1992), p. 30.

27 "fair fugitive . . .": Knight (1992), p. 32.

27 "Davy has actually invented . . .": Fullmer (2000), p. 221.

28 "Sir, I have read with admiration . . .": Joseph Priestley to Humphry Davy, October 31, 1801; Davy (1858), p. 51.

28 "a miraculous young man . . .": Hartley (1966), p. 22.

29	"delight was in his intellectual being . . .": Forgan (1980), p. 177.
29	"Why, Davy could eat them all . . .": Hartley (1966), p. 23.
29	"the application of science . . .": James (2000), p. 2.
30	"grand theatre": Fullmer (2000), p. 344.
30	"The sensation created by his first course . . .": Treneer (1963), p. 86.
30	"it was impossible to have seen him . . .": Fullmer (2000), p. 344.
30	"Those eyes were made for something . . .": Williams (1965), p. 19.
31	"Not contented with what is found . . .": Davy (1839–1840), p. 318.
31	No wonder Coleridge attended Davy's lectures . . . : Hartley (1966), p. 45.
31	"guardians of civilization . . ."; "friends and protectors . . .": Davy (1839–1840), p. 323.
32	"With all his immense ability . . .": Thompson (1898), p. 32.
33	"water, chemically pure, is decomposed by electricity . . .": Paris (1831), p. 151.
36	"[I] am now working at my old trade . . .": Faraday to Huxtable, October 18, 1812; Williams (1960b), p. 528.
38	"I am constantly engaged . . .": Faraday to an unidentified aunt and uncle, September 13, 1813; James (1991–1996), vol. 1, p. 66.
38	"If we never judge . . .": Faraday to Benjamin Abbott, June 1, 1813; James (1991–1996), vol. 1, p. 55.
38	"the generality of mankind . . .": Faraday to Benjamin Abbott, June 11, 1813; James (1991–1996), vol. 1, p. 60.
39	"As when on some secluded branch . . ." Faraday to Benjamin Abbott, June 18, 1813; James (1991–1996), vol. 1, p. 63.
39	"[T]he music is so excellent . . .": Faraday to Benjamin Abbott, May 12, 1813; BJI, p. 64.
39	". . . I keep regular hours . . .": Faraday to Benjamin Abbott, October 11, 1812; James (1991–1996), vol. 1, p. 39.
39	"little frolics": James (1992b), p. 237.
39	"What is the longest . . .": Faraday to Benjamin Abbott, August 2, 1812; James (1991–1996), vol. 1, p. 12.
40	"–no–no–no–no . . .": Faraday to Benjamin Abbott, October 1, 1812; James (1991–1996), vol. 1, p. 36.

40 "What a singular compound is man . . .": Faraday to Benjamin Abbott, May 14, 1813; James (1991–1996), vol. 1, p. 53.

42 " 'Tis indeed a strange venture . . .": Faraday's Journal, October 13, 1813; Bowers and Symons (1991), p. 1.

42 "This morning formed a new epoch in my life.": Faraday's Journal, October 13, 1813; Bowers and Symons (1991), p. 1.

CHAPTER 3: THE UNIVERSITY OF EXPERIENCE

44 "I was more taken by the scenery . . .": Faraday's Journal, October 15, 1813; Bowers and Symons (1991), p. 2.

44 "When Sir H. Davy first had the goodness . . .": Faraday to his mother, April 14, 1814; Bowers and Symons (1991), p. 82.

44 "I could hardly help laughing . . .": Faraday's Journal, October 19, 1813; Bowers and Symons (1991), p. 4.

45 "horses pigs poultry . . ."; "I think it is impossible . . .": Faraday's Journal, October 21, 1813; Bowers and Symons (1991), p. 7.

45 "I am quite out of patience . . .": Faraday's Journal, December 21, 1813; Bowers and Symons (1991), p. 33.

45 "I know nothing of the language . . .": Faraday's Journal, October 29, 1813; Bowers and Symons (1991), p. 15.

45 "[E]very day presents sufficient to fill a book . . .": Faraday to Benjamin Abbott, May 1, 1814; Bowers and Symons (1991), p. xiv.

46 "To attempt to decompose Nitrogen . . .": Hartley (1966), p. 106.

46 "both the Glory and the disgrace . . ."; "a nation of thieves": Faraday's Journal, December 30, 1813; Bowers and Symons (1991), p. 15.

46 "He was sitting in one corner of his carriage . . .": Faraday's Journal, December 19, 1813; Bowers and Symons (1991), p. 33.

46 "I have found the French people . . .": Faraday to Robert G. Abbott, August 6, 1814; James (1991–1996), vol. 1, p. 80.

47 "At Paris civilization has been employed . . .": Faraday to Robert G. Abbott, August 6, 1814; Bowers and Symons (1991), p. 118.

47 "as a stranger who had not always opportunities": Faraday's Journal, November 9, 1813; Bowers and Symons (1991), p. 18.

47 "[T]he constant presence of Sir Humphry Davy . . .": Faraday to Robert G. Abbott, August 6, 1814; Bowers and Symons (1991), p. 116.

49 "The French Chemists were not aware . . .": Faraday to Benjamin Abbott, February 23, 1815; James (1991), vol. 1, p. 128.

49 "[A]s is the practice with him . . .": Faraday to T. Huxtable, February 15, 1815; James (1991–1996), vol. 1, p. 125.

49 "I do not think I ever saw a more beautiful scene . . .": Faraday's Journal, December 29, 1813; Bowers and Symons (1991), p. 34.

49 "Nearer I behold . . .": Knight (1992), p. 101.

50 " 'Tis a pleasant state . . .": Faraday's Journal, December 30, 1813; Bowers and Symons (1991), p. 39.

50 "It has a degree of positive excellence . . .": Faraday's Journal, January 15, 1814; Bowers and Symons (1991), p. 41.

52 "Tell B[enjamin Abbott] I have crossed the Alps . . .": Faraday to his mother, April 14, 1814; Bowers and Symons (1991), p. 83.

52 "Whenever a vacant hour occurs . . .": Faraday to his mother, April 14, 1814; Bowers and Symons (1991), p. 82.

52 "I have no scientific companion[s] . . .": Benjamin Abbott to Faraday, November 20–22, 1814; James (1991), vol. 1, p. 94.

53 "almost to adoration": Faraday to his mother, April 14, 1814; Bowers and Symons (1991), p. 84.

53 "The civilization of Italy . . .": Faraday to Robert G. Abbott, August 6, 1814; James (1991–1996), vol. 1, p. 81.

53 "We admired Davy . . .": Gooding and James (1989), p. 40.

53 "And now, my dear Sirs . . .": Bowers and Symons (1991), p. xi.

54 "Alas! how foolish perhaps was I . . .": Faraday to Benjamin Abbott, November 26, 1814; Williams (1965), p. 39.

54 "She is haughty and proud . . .": Faraday to Benjamin Abbott, January 25, 1815; Williams (1965), p. 40.

55 "At present I laugh at her whims . . .": Faraday to Benjamin Abbott, January 25, 1815; James (1991–1996), vol. 1, p. 117.

55 "was an oversight, if not a delusion . . .": Knight (1992), p. 92.

55 "licentiousness and riot": Faraday to his sister Elizabeth, December 21, 1814; James (1991–1996), vol. 1, p. 103.

55 "Rome glittered with Princes . . .": Faraday to Benjamin Abbott, January 25, 1815; Bowers and Symons (1991), p. 151.

55 "As for me, like a poor unmanned, unguided skiff . . .": Faraday to T. Huxtable, February 13, 1815; James (1991–1996), vol. 1, p. 124.

56 "It is with no small pleasure I write you . . .": Faraday to his mother, April 16, 1815; James (1991), vol. 1, p. 128.

CHAPTER 4: FEAR AND CONFIDENCE

58 "She has a temper . . .": Treneer (1963), p. 177.

58 "accuracy and steadiness of manipulation": Williams (1965), p. 45.

58 "Pray make an investigation of this subject . . .": Davy to Faraday, August 3, 1815; James (1991–1996), vol. 1, p. 133.

58 "Mr. Hatchett's letter contained praises . . .": Davy to Faraday, postmarked October 29 1818; James (1991–1996), vol. 1, p. 172.

58 "It gives me great pleasure . . .": Davy to Faraday, May 15, 1819; James (1991–1996), vol. 1, p. 181.

59 "was never brilliant or eloquent . . .": Hamilton (2002), p. 40.

59 "I have uniformly received . . .": Hamilton (2002), p. 143.

59 "from the large running letters of Brande . . .": Gladstone (1873), p. 16.

60 "We have subdued this monster.": Treneer (1963), p. 167.

60 "I was witness in our laboratory . . .": Treneer (1963), p. 174.

61 "The opinion of the inventor.": Thompson (1898), p. 43.

61 "when my fear was greater than my confidence . . .": Williams (1965), p. 44.

61 "If M. Davy would be so kind . . .": Williams (1965), p. 46.

62 "[W]e are all liable to error . . .": Day (1999), p. 103.

62 "I am continually saying to myself . . .": Faraday to William Flexman, May 3, 1818; James (1991–1996), vol. 1, p. 161.

64 ". . . only take care you do not kill yourself . . .": Faraday to William Flexman, May 3, 1818; James (1991–1996), vol. 1, p. 162.

64 "I would, if possible, imitate a tree . . .": Faraday to Benjamin Abbott, December 31, 1816; James (1991–1996), vol. 1, p. 149.

64 His powers, unshackled . . . : Gladstone (1873), p. 11.

65 "The [natural] philosopher should be a man . . .": BJI, p. 220.

65 "[The satisfaction] resulting from judgement . . .": "On Imagination and Judgement," from *A Class Book for the Reception of Mental Exercises Instituted July 1818,* pp. 24–25; quoted in Williams (1965), p. 81.

65 "A sensitive mind . . .": "On the Pleasures and Uses of the Imagination," from *A Class Book for the Reception of Mental Exercises Instituted July 1818,* p. 41; quoted in Williams (1965), p. 82.

66 "Democratical–comical trade . . .": Faraday's *Common Place Book,* 1816–; quoted in Berman (1978).

66 "What is the pest and plague of human life . . .": Faraday's *Common Place Book,* 1816–; quoted in Williams (1965), p. 96.

67 "You know me as well . . .": Faraday to Sarah Barnard, July 5, 1820; James (1991–1996), vol. 1, p. 199.

68 "I wished for a moment . . .": BJI, p. 318.

68 The cliffs rose like mountains . . .: BJI, pp. 319–320.

69 My dear Sarah,–It is astonishing . . .: Faraday to Sarah Barnard, December 1820; James (1991–1996), vol. 1, p. 202.

70 "There will be no bustle . . .": Faraday to Mary Reid, June 1821; James (1991–1996), vol. 1, p. 209.

70 "quite content to be the pillow . . .": Williams (1965), p. 99.

71 "I am tired of the dull detail of things . . .": Faraday to his wife Sarah, July 21, 1822; Day (1999), p. 21.

71 "That is between me and my God.": Williams (1965), p. 104.

CHAPTER 5: RISING TO THE LIGHT

74 "You are quite right to say that it is inconceivable . . .": Williams (1965), pp. 142–143.

76 "All the phenomena presented . . .": Williams (1965), p. 144.

77 "I have really been ashamed . . .": Faraday to Gaspard de la Rive, October 9, 1822; James (1991–1996), vol. 1, p. 292.

78 Nothing is more difficult . . . : Williams (1965), p. 89.

78 "I am naturally sceptical in the matter of theories . . .": Faraday to André-Marie Ampère, February 2, 1822; Williams (1965), p. 168.

80 "There they go . . .": Thompson (1898), p. 51.

80 "I shall never forget the enthusiasm . . .": BJI, p. 345.

81 "Very satisfactory . . .": Gooding and James (1989), p. 120.

82 "I hear every day . . .": Faraday to James Stodart, October 8, 1821; James (1991–1996), vol. 1, p. 228.

83 "I am bold enough sir to beg . . .": Faraday to William Hyde Wollaston, October 30, 1821; James (1991–1996), vol. 1, p. 233.

83 "You seem to me to labour . . .": William Hyde Wollaston to Faraday, November 1, 1821; James (1991–1996), vol. 1, p. 235.

84 "Had not an experiment on the subject . . .": BJI, p. 346.

84 "I am compelled to say I have not found . . .": Faraday to André-Marie Ampère, July 1, 1823; James (1991–1996), vol. 1, p. 321.

85 "unaided by any knowledge . . .": Hamilton (2002), p. 189.

85 "I spoze you noze as I did your bizzness . . .": Richard Phillips to Faraday, May 3, 1823; James (1991–1996), vol. 1, p. 315.

86 "Sir H. Davy angry . . .": Faraday to Henry Warburton, May 30, 1823; James (1991–1996), vol. 1, p. 318.

87 "Sir H. Davy told me I must . . .": Williams (1965), p. 160.

87 "You would probably find yourself engaged . . .": Faraday to Henry Warburton, May 30, 1823; James (1991–1996), vol. 1, p. 318.

87 "When I meet with any of those . . .": Henry Warburton to Faraday, July 8, 1823; James (1991–1996), vol. 1, p. 323.

88 "sought and paid for": MacDonald (1964), p. 25.

88 "health and success": Humphry Davy to Faraday, June 29, 1823; James (1991–1996), vol. 1, p. 320.

88 "Brothers in intellect . . .": Tyndall (2001), p. 396.

89 The mighty birds still upward rose . . .: Treneer (1963), p. 203.

89 "A father is not always wise . . .": Tyndall (2001), p. 395.

89 "I send you in the original . . .": Faraday to John Ayrton Paris, December 23, 1829; James (1991–1996), vol. 1, p. 497.

90 "I never yet even in my short time . . .": Faraday to Ampère, November 17, 1825; James (1991–1996), vol. 1, p. 392.

90 "Well, but, Tyndall, I *am* humble . . .": Tyndall (1961), p. 46.

90 "We have heard much of Faraday's gentleness . . .": Tyndall (1961), pp. 44–45, including Proverbs, 16:32.

91 "I was by no means in the same relation . . .": BJI, p. 353.

91 "a model to teach him . . .": BJI, p. 210.

CHAPTER 6: HE SMELLS THE TRUTH

94 "Every letter you write me . . .": Faraday to Ampère, November 17, 1825; James (1991–1996), vol. 1, p. 392.

96 "Evening opportunities–interesting, amusing . . .": BJI, p. 392.

96 "I am persuaded that all persons . . .": Faraday, "Observations on Mental Education," May 6, 1854; Williams (1965), p. 337.

96 "praise heartily . . . censure mildly": Hamilton (2002), p. 152.

96 "one of the most successful chemical enquirers . . ."; "This meritorious and valuable institution . . .": Hamilton (2002), p. 202.

97 "The *laws of nature,* as we understand them . . .": Faraday, "Observations on Mental Education," May 6, 1854; Williams (1965), p. 336.

98 "first instructress": Hamilton (2002), p. 220.

98 "He speaks with an ease and freedom . . .": Hamilton (2002), p. 208.

98 "It was an irresistible eloquence . . .": Williams (1965), p. 333.

98 "All is a sparkling stream of eloquence . . .": Seeger (1968), p. 35.

98 "[H]is Friday evening discourses . . .": Gooding and James (1989), p. 63.

99 "No attentive listener ever came away . . .": Gooding and James (1989), p. 63.

99 A flame should be lighted . . .: Faraday to Benjamin Abbott, June 1, 4–5, 11, and 18, 1813; James (1991), vol. 1, pp. 55–65; Hamilton (2002), p. 201.

100 "[H]ow often have I felt oppression . . .": Faraday to Benjamin Abbot, June 1, 1813; James (1991–1996), vol. 1, p. 56.

101 "I will return to second childhood . . .": Faraday (1993), p. 5.

101 "Look at these colours . . .": Seeger (1968), p. 32.

101 "Always remember that when a result happens . . .": Seeger (1968), p. 32.

102 "[I]n the pursuit of science we first start . . .": Seeger (1968), p. 33.

102 "We young ones have a perfect right . . .": Faraday (1960), p. 72.

102 "All I can say to you at the end . . .": Faraday (1960), p. 97.

103 "Let us now consider, for a little while . . .": Faraday (1993), p. 5.

103 "He made us all laugh heartily . . .": Thompson (1898), p. 237.

103 [W]hat study is there more fitted to the mind . . .: Faraday (1993), p. 88.

104 "At the end of the lectures . . .": Yarrow (1926–28), p. 480.

104 "After I went, in 1826, to stay . . .": BJI, p. 421.

105 "How I rejoiced to be allowed to go . . .": BJI, pp. 421–424.

106 "Nothing vexed him more . . .": BJI, p. 423.

106 "In times of grief or distress . . .": BJI, p. 424.

106 I have been watching the clouds . . . : Faraday to his brother-in-law Edward Barnard, July 23, 1826; James (1991–1996), vol. 1, pp. 419–420.

109 "united vast strength with perfect flexibility . . .": Tyndall (1961), p. 23.

109 "Er riecht die Wahrheit": Tyndall (1961), p. 54.

109 "Papers of mine published . . .": BJII, p. 55.

109 "nervous headaches and weakness": Faraday to Richard Phillips, August 29, 1828; James (1991–1996), vol. 1, p. 465.

110 "a perfect piece of glass . . .": BJI, p. 401.

110 "I further wish you most distinctly to understand . . .": Faraday to Davies Gilbert, president of the Royal Society, May 13, 1830; James (1991–1996), vol. 1, pp. 517–518.

110 "I had enough of endeavouring . . .": Faraday to John James Chapman, officer in the Royal Artillery, August 11, 1832; James (1991–1996), vol. 2, p. 75.

CHAPTER 7: A TWITCH OF THE NEEDLE

111 "Michael Faraday"; "Mr. Professor Faraday . . .": Gooding and James (1989), p. 58.

111 "The [Royal] Institution has been a source of knowledge . . .": Faraday to Dioysius Lardner, Professor of Natural Philosophy and Astronomy, University of London, October 6, 1827; James (1991–1996), vol. 1, p. 442.

112 "[M]y time is my only estate . . .": Faraday to Percy Drummond, lieutenant governor of the Royal Military Academy, Woolrich, June 29, 1928; James (1991–1996), vol. 1, p. 487.

113 "Expts. on the production of Electricity . . .": Faraday (1932–1936), vol. I, p. 367.

118 "I am inclined to compare . . .": Williams (1965), p. 181.

119 "I am busy just now again . . .": Faraday to Richard Phillips, September 23, 1831; James (1991–1996), vol. 1, pp. 579–580.

120 "a mere momentary push or pull . . .": Faraday (1932–1936), p. 372.

120 "powerful effect . . . pulling the needle quite round": Faraday (1932–1936), p. 373.

120 "*Indeed* I am sorry for it . . .": Tyndall (1961), p. 39.

123 "I know not, but I wager that one day . . .": Williams (1965), p. 196.

124 "I have rather . . . been desirous of discovering . . .": ERE (1832), vol. 1, series 2, p. 47, ¶ 159.

CHAPTER 8: TOIL AND PLEASURE

125 "We are here to refresh . . ."; "It is quite comfortable to me . . ."; "Excuse this egotistical letter . . .": Faraday to Richard Phillips, November 29, 1831; James (1991–1996), vol. 1, pp. 590–591.

126 "overran in a single autumn . . .": Tyndall (1961), p. 40.

126 "I never took more pains . . .": Faraday to William Jerden, March 37, 1832; James (1991–1996), vol. 2, p. 29.

127 "To render you complete justice . . .": J. N. P. Hachette to Faraday, May 18, 1832; James (1991–1996), vol. 2, p. 50.

129 "a wire at rest in the neighborhood . . .": ERE (1832), vol. 1, series 2, p. 69, ¶ 242.

130 "He hated what he called . . .": Tyndall (1961), p. 49.

132 "By current, I mean anything progressive . . .": ERE (1833), vol. 1, series 3, p. 81, ¶ 283.

135 "I had some hot objections . . .": Faraday to William Whewell, May 15, 1834; James (1991–1996), vol. 2, p. 186.

136 "All tends to prove that chemical affinity . . .": BJII, p. 69.

137 "My matter . . . overflows . . .": Faraday to John William Lubbock, November 2, 1833; James (1991–1996), vol. 2, pp. 154–155.

CHAPTER 9: A CAGE OF HIS OWN

138 "I had begun to imagine that I thought . . .": Faraday to William Whewell, September 19, 1835; James (1991–1996), vol. 2, p. 277.

142 "I went into the cube and lived in it . . .": ERE (1837), vol. 1, series 11, p. 366, ¶ 1174.

143 "Gravity . . . will not turn a corner": Tyndall (1961), p. 82.

144 "The science of electricity is in that state . . .": ERE (1837), vol. 2, series 11, p. 360, ¶ 1161.

144 "[I]n whatever way I view it . . .": ERE (1837), vol. 2, series 11, p. 386, ¶ 1231.

145 "I am unfortunate in a want of mathematical knowledge . . .": Faraday to Ampère, September 3, 1822; James (1991–1996), vol. 1, p. 287.

145 "I have far more confidence in the one man . . .": Faraday to John Tyndall, April 19, 1851; James (1991–1996), vol. 4, p. 281.

146 "If Newton, not without reason, has been compared . . .": Peter Riess to Faraday, August 9, 1855; James (1991–1996), vol. 4, p. 892.

146 "The salient quality of Faraday's scientific character . . .": Tyndall (2001), p. 401.

147 "so treacherous . . . remember . . .": Faraday to Christian Friedrich Schoenbein, May 16, 1843; Hare (1976), p. 36.

147 "I have been convinced by long experience . . .": Faraday to George Biddell Airy, March 10, 1832; James (1991–1996), vol. 2, p. 24.

147 "I am greatly obliged to you . . .": Faraday to John Britton, February 29, 1832; James (1991–1996), vol. 2, p. 23.

147 ". . . am under the necessity of declining it . . .": Faraday to John Rennie, February 26, 1835; James (1991–1996), vol. 2, p. 236.

148 ". . . now several years since I have dined . . .": Faraday to William Charles Macready, March 27, 1838; James (1991–1996), vol. 2, p. 493.

148 ". . . am so circumstanced that I cannot possibly avail myself . . .": Faraday to Thomas Phillipps, June 15, 1838; James (1991–1996), vol. 2, p. 508.

148 ". . . quite shut out from the enjoyment . . .": Faraday to George Biddell Airy, June 2, 1840; James (1991–1996), vol. 2, p. 674.

148 "Mr. Faraday is very grateful to Miss Coutts . . .": Faraday to Angela Georgina Burdett Coutts, August 15, 1839; James (1991–1996), vol. 2, p. 600.

148 "I called on Faraday this morning . . .": Charles Wheatstone to W. F. Cooke, October 4, 1838; Thomas (1991), p. 124.

148 "called to see Mr. Faraday . . .": James (1992b), p. 234.

148 "continual calls upon his time and thought . . .": BJII, p. 223.

148 "I have been here so long . . .": Faraday to John Millington, former professor of mechanics at the Royal Institution, November 12, 1836; James (1991–1996), vol. 2, p. 386.

149 "He confessed to me . . .": Hare (1976), pp. 34–35.

149 "He looks up to this work . . .": Hare (1976), p. 35.

149 "My medical friends have required me to lie by . . .": Faraday to Christian Friedrich Schoenbein, March 27, 1841; James (1991–1996), vol. 3, p. 11.

CHAPTER 10: AN EXCELLENT DAY'S WORK

150 "[T]his is to declare in the present instance . . .": Thompson (1898), p. 223.

150 "You know that I am a recluse & unsocial . . .": Faraday to Jean-Baptiste-André Dumas, April 29, 1840; James (1991–1996), vol. 2, p. 657.

151 "low nervous attacks"; "memory so treacherous . . ."; "disobedient to the will": Faraday to Christian Friedrich Schoenbein, May 16, 1843; James (1991–1996), vol. 3, p. 146.

151 "Mrs. Faraday drew me aside . . .": Crosse (1891), pp. 34–35.

152 "Mr Faraday seems very unwilling to write letters . . .": Sarah Faraday to Edward Magrath, August 14, 1841; James (1991–1996), vol. 3, pp. 32–33.

152 "Now as to the main point of this trip . . .": Faraday to Edward Magrath, August 15, 1841; James (1991–1996), vol. 3, p. 34.

153 "I would gladly give half my strength . . .": BJII, p. 142.

153 "You have all the confidence of unbaulked health . . .": Faraday to Ada Lovelace, October 24, 1844; James (1991–1996), vol. 3, p. 265.

154 "I have long been vowed to the Temple . . .": Ada Lovelace to Faraday, October 16, 1844; James (1991–1996), vol. 3, p. 253.

154 "I have been reading . . . of Faraday's researches . . .": James (1991–1996), vol. 3, p. xxxvi.

155 "inoculated with Faraday fire": Berkson (1974), p. 136.

155 "I purpose resuming this subject hereafter . . .": Faraday to William Thomson, August 8, 1845; James (1991–1996), vol. 3, p. 407.

157 "terribly powerful": Diary, August 30, 1845, vol. 4, p. 258, ¶ 7457.

158 "This fact will most likely prove exceedingly fertile . . .": Diary, September 13, 1845, vol. 4, p. 264, ¶ 7504.

158 "Have got enough for today": Diary, September 13, 1845, vol. 4, p. 267, ¶ 7536.

158 *"An excellent day's work"*: Diary, September 18, 1845, vol. 4, p. 277, ¶ 7610.

159 "I believe that, in the experiments . . .": ERE (1845), vol. 3, series 19, p. 1, introductory footnote.

159 "At present, I have scarcely a moment . . .": Faraday to Christian Friedrich Schoenbein, November 13, 1845; James (1991–1996), vol. 3, p. 428.

159 "I have had your letter by me on my desk . . .": Faraday to Auguste de la Rive, December 4, 1845; James (1991–1996), vol. 3, pp. 438–439.

160 "It is curious to see such a list as this . . .": ERE (1845), vol. 3, series 20, p. 36, ¶ 2281.

CHAPTER 11: NOTHING IS TOO WONDERFUL TO BE TRUE

162 "I must remain plain Michael Faraday . . .": Tyndall (1961), p. 186.

162 "The atomic doctrine . . . is not so carefully distinguished . . .": ERE (1844), vol. 2, p. 285.

162 "[A]ll our perception and knowledge of the atom . . .": ERE (1844), vol. 2, pp. 290–291.

163 "[T]hat which represents size may be considered . . .": ERE (1846), vol. 3, p. 447.

163 "The powers around the centres . . .": ERE (1844), vol. 2, p. 291.

164 "tosses the atomic theory from horn to horn . . .": Tyndall (1961), p. 148.

164 "an atmosphere of force . . .": Williams (1965), p. 378.

164 "I cannot doubt but that he who . . .": ERE (1844), vol. 2, pp. 285–286.

165 "All the views which Faraday has brought forward . . .": William Thomson, "On the Mathematical Theory of Electricity in Equilibrium," *Cambridge and Dublin Mathematical Journal,* November 1845; Williams (1965), p. 513.

165 "one of the most singular speculations...": Tyndall (1961), p. 149.

167 "The view which I am so bold as to put forth...": ERE (1846), vol. 3, p. 451.

167 "[F]rom first to last, understand...": ERE (1846), vol. 3, p. 447.

167 "The propagation of light...": ERE (1846), vol. 3, p. 451.

168 "I think it is likely that I have made many mistakes...": ERE (1846), vol. 3, p. 452.

168 "It is amusing to see how many write..."; "Faraday's achievements are due to his immense earnestness...": John Tyndall to Thomas Hirst, November 5, 1855; Williams (1965), p. 509.

169 "shake men's minds from their habitual trust"; "[I]t is better to be aware, or even to suspect...": ERE (1854), vol. 3, p. 565.

169 "Do not think learning in general is arrived at its perfection...": Watts (1809), p. 20.

170 "That gravity should be innate, inherent and essential...": ERE (1853), vol. 3, p. 507.

171 "At present we are accustomed to admit action...": ERE (1855), vol. 3, p. 570.

171 "It is, probably, of great importance...": ERE (1855), vol. 3, p. 570.

171 "I think it will be exceedingly difficult...": ERE (1855), vol. 3, pp. 571–572.

171 "...the power is always existing around the sun...": ERE (1855), vol. 3, p. 574.

172 "I can hardly imagine anyone...": George Biddell Airy to the Reverend John Barlow, February 7, 1855; BJII, p. 353.

172 "How few understand the physical lines of force!" Diary of Margery Ann Reid, Faraday's niece, November 7, 1855; Williams (1965), p. 507.

172 "The lines of force... stood before his intellectual eye...": Watson (1957), p. 338.

173 "Nothing is too wonderful to be true...": Diary, March 19, 1849, Vol. 5, p. 152, ¶ 10040.

Chapter 12: A Partaker of Infinity

175 "What's the go o' that?": C&G, p. 16.

175 "He is a very happy man . . .": C&G, pp. 15–16.

175 "frog-like over a handkerchief . . .": Goldman (1983), p. 34.

175 "was at first regarded as shy and rather dull . . .": C&G, p. 62.

175 "They never understood me . . .": Tolstoy (1981), p. 20.

176 "he surprised his companions . . .": C&G, p. 62.

176 "While extremely neat in person . . .": C&G, p. 64.

177 "Jamsie, you're in a prop": C&G, p. 64.

177 "unquestionably the most extraordinary man . . .": C&G, p. 88.

177 "When I waken I do so either at 5.45 or 9.15 . . .": Maxwell to Lewis Campbell, September 22, 1848; C&G, p. 77.

177 "I have set up shop now above the wash-house . . .": Maxwell to Lewis Campbell, July 5–6, 1848; C&G, pp. 74–75.

179 "matter in the wrong place"; "Jamsie's dirt": C&G, pp. 100, 103.

179 "Objectivity alone is favourable . . .": Maxwell to Lewis Campbell, February 10, 1852; C&G, p. 125.

179 "stimulate the circulation": C&G, p. 115.

180 "Now, my great plan, which was conceived of old . . .": Maxwell to Lewis Campbell, March 7, 1852; C&G, p. 126.

180 "Happy is the man who can recognize . . .": C&G, pp. 144–145.

181 "There is scarcely a single topic that he touched upon . . .": Goldman (1983), p. 176.

181 "most remarkable applications of Mathematics to physics . . .": Goldman (1983), p. 71.

181 "old traditions about forces acting at a distance": Maxwell to Faraday, October 19, 1861; C&G, p. 245.

181 "a first step in right thinking": Maxwell to Faraday, October 19, 1861; C&G, p. 245.

181 "The method which Faraday employed . . .": Maxwell (1904), vol. II, p. 176.

182 "It was perhaps for the advantage of science . . .": Maxwell (1904), vol. II, p. 176.

182 "I do not think that it contains even the shadow . . .": Goldman (1983), p. 141.

183 "It was like listening to a fairy-tale": C&G, p. 143.

CHAPTER 13: THE LIGHT UNSEEN

184 "I was at first almost frightened . . .": Faraday to Maxwell, March 25, 1857; Faraday (1971), p. 864.

184 "The present state of electrical science . . .": Maxwell (1856), p. 27.

185 "to be of an indefinite and unmathematical character": Maxwell (1856), p. 29.

185 "By the method which I adopt . . .": Maxwell (1856), p. 29.

185 "[I]f I could live my life over again . . .": Crosse (1891), p. 37.

185 "I received your paper, and thank you very much for it . . .": Faraday to Maxwell, March 25, 1857; Faraday (1971), p. 864.

186 "First—the quickness of the action . . .": Berkson (1974), p. 110.

186 " 'weave a web across the sky' . . .": Maxwell to Faraday, November 9, 1857; C&G, p. 203.

186 "If on a former occasion I seemed to ask you . . .": Faraday to Maxwell, November 13, 1857; C&G, pp. 205–206.

187 "There is one thing I would be glad to ask you . . .": Faraday to Maxwell, November 13, 1857; C&G, p. 206.

187 "What a painful amount of modesty he has . . .": Maxwell to William Thomson, November 14, 1857; Goldman (1983), p. 146.

187 "Maxwell's lectures were, as a rule, most carefully arranged . . .": Goldman (1983), p. 76.

188 "Ho, Maxwell, cannot you get out?": Goldman (1983), p. 92.

189 "I think we have now strong reason to believe . . .": Maxwell to Faraday, October 19, 1861; C&G, p. 244.

189 "The electromagnetic theory of light, as proposed . . .": James (1998), p. 80.

190 "Maxwell's theory is Maxwell's equations": Simpson (2001), p. 384.

190 "Today, we understand better that what counts . . .": Feynman (1964), vol. 1, chap. 18, p. 2.

191 "It is clear that the experiments amount to so many reasons . . .": Glashow (1994), p. 360.

191 "I would recommend the student . . .": Maxwell (1904), p. xi.

CHAPTER 14: THE SIMPLEST EARTHLY PLACE

195 "rolled up in clouds so dense . . .": Faraday to *The Times,* July 7, 1855, printed July 9, 1855; James (1991–1996), vol. 4, p. 882.

195 "sweeter atmosphere"; "If we neglect this subject . . .": Faraday to *The Times,* July 7, 1855, printed July 9, 1855; James (1991–1996), vol. 4, p. 882.

196 "And we hope the Dirty Fellow will consult . . .": James (1991–1996), vol. 4, p. 883.

196 "diabolical or supernatural agency": Faraday to *The Times,* June 28, 1853; James (1991–1996), vol. 4, pp. 525–526.

196 "Shall we educate ourselves . . .": Seeger (1968), p. 36.

197 "never seen anything incompatible . . .": BJII, pp. 325–326.

197 "unclean spirits . . . waking in the hearts of men": Faraday to Caroline Deacon, July 23, 1853; James (1991–1996), vol. 4, p. 539.

197 "will be ranked among the most astonishing discoveries . . .": "appearing to increase equably . . .": John Allen to Faraday, May 16, 1853; James (1991–1996), vol. 4, p. 513.

197 "Are we not on the eve of some new discovery . . .": "Returning home I called my friend . . .": William Edward Hickson to Faraday, May 17, 1853; James (1991–1996), vol. 4, pp. 515–516.

198 "produce all the phenomena . . .": William Edward Hickson to Faraday, May 22, 1853; James (1991–1996), vol. 4, p. 518.

198 "not a showman, but simply a bankers' clerk": G. G. Wilson to Faraday, October 26, 1853; James (1991–1996), vol. 4, p. 585.

198 "[T]he world is running mad . . .": Faraday to Caroline Deacon, Faraday's niece, July 23, 1853; James (1991–1996), vol. 4, pp. 538–539.

198 "I declare that taking the average of many minds . . .": Faraday to Christian Friedrich Schoenbein, July 25, 1853; James (1991–1996), vol. 4, p. 542.

198 "I do not object to table-moving . . .": Faraday, "Observations on Mental Education," May 6, 1854; Williams (1965), p. 337.

199 Study science with earnestness . . . : BJII, pp. 309–310.

199 "turning the tables upon the table turners": Faraday to Christian Friedrich Schoenbein, July 25, 1853; James (1991), vol. 4, p. 542.

200 "If the hands involuntarily moved...": Faraday to *The Times,* June 28, 1853; James (1991–1996), vol. 4, pp. 526–527.

200 "Sir,–I beg to thank you...": Faraday to Thomas S–, Esq., November 1, 1864; BJII, p. 468.

200 "I have been greatly startled...": Faraday to *The Times,* June 28, 1853; James (1991–1996), vol. 4, p. 527.

201 "education has for its first and its last step...": Faraday, "Observations on Mental Education," May 6, 1854; Williams (1965), p. 337.

201 "It is necessary that a man *examine himself*...": Faraday, "Observations on Mental Education," May 6, 1854; Williams (1965), p. 338.

201 "The development of the applications...": Williams (1965), p. 340.

202 "That the natural knowledge that has been given...": BJII, pp. 453–454.

202 "[T]hey come to me and they talk...": Hamilton (2002), p. 391.

203 "honour and encourage natural knowledge": Williams (1965), p. 343.

203 "I am in deep distress...": Faraday to William Buchanan, November 3, 1850; James (1991–1996), vol. 4, p. 196.

204 "I rise from a restless bed...": Sarah Faraday to William Buchanan, October 31, 1850; James (1991–1996), vol. 4, p. 195.

204 "God has wrought with me...": Faraday to William Buchanan, November 6, 1850; James (1991–1996), vol. 4, p. 200.

204 "My memory wearies me greatly...": Faraday to the Reverend John Barlow, August 19, 1857; Faraday (1971), p. 878.

205 "Surely the force of gravitation...": Diary, February 10, 1859, vol. 7, pp. 334–337, ¶ 15785–15809.

205 "I went to Dover last Monday...": Williams (1965), p. 491.

206 "At two o'clock he came down for me...": Tyndall (2001), pp. 406–407.

206 "[Faraday] is as simple, charming, and unaffected...": James (1991–1996), vol. 4, p. xliii.

207 It is with the deepest feeling that I address you... : Williams (1965), pp. 497–498.

208 Personal explanation,–years of happiness here... : Williams (1965), p. 498.

209 "When I try to write [to you] . . .": Faraday to Christian Friedrich Schoenbein, March 27, 1860; BJII, p. 438.

209 "Again and again I tear up my letters . . .": Faraday to Christian Friedrich Schoenbein, September 1862; BJII, p. 455.

209 "My head is full, and my heart also . . .": Faraday to Sarah Faraday, August 14, 1863; BJII, p. 458.

209 "altogether a very tottering and helpless thing": Faraday to Dr. Holzmann, December 22, 1863; BJII, p. 465.

210 "His mind was then failing . . .": Williams (1965), p. 500.

210 "have a plain simple funeral . . .": Faraday to Sir James South, January 1866; BJII, p. 478.

210 "came out from the shrubbery": Gladstone (1873), p. 59.

EPILOGUE

212 "greatest alteration . . . in our conception . . .": Friedel (1981).

212 "Oh, to live always in it!": Thompson (1898), p. 240.

BIBLIOGRAPHY

Agassi, Joseph. "An Unpublished Paper of the Young Faraday." *Isis* 52 (1961): 87–90.

——. *Faraday as a Natural Philosopher.* Chicago: University of Chicago Press, 1971.

Berkson, William. *Fields of Force: The Development of a World View from Faraday to Einstein.* New York: John Wiley & Sons, 1974.

Berman, Morris. *Social Change and Scientific Organization: The Royal Institution, 1799–1844.* Ithaca, NY: Cornell University Press, 1978.

Bernal, J. D. *A History of Classical Physics from Antiquity to the Quantum.* New York: Barnes & Noble Books, 1997.

Bowers, Brian, and Lenore Symons. *Curiosity Perfectly Satisfied: Faraday's Travels in Europe, 1813–1815.* London: Peter Peregrinus Ltd. in association with the Science Museum, London, 1991. [Contains Faraday's journal of the trip.]

Bragg, William Henry. "Michael Faraday." *Scientific Monthly* 33 (1931): 481–499.

Butterfield, Herbert. *The Origins of Modern Science 1300–1800.* New York: Free Press, 1965.

Campbell, Lewis, and William Garnett. *The Life of James Clerk Maxwell, with Selections from His Correspondence and Occasional Writings.* London: Macmillan and Co., 1884. [**C&G**]

Cantor, Geoffrey. "Why Was Faraday Excluded From the Sandemanians in 1844?" *British Journal for the History of Science* 22 (1989): 433–437.

——. *Michael Faraday: Sandemanian and Scientist.* London: Macmillan, 1991.

———. "The Scientist as Hero: Public Images of Michael Faraday." In *Telling Lives in Science: Essays on Scientific Biography,* edited by Michael Shortland and Richard Yeo. Cambridge, UK: Cambridge University Press, 1996.

Cantor, Geoffrey, David Gooding, and Frank A. J. L. James. *Faraday.* Atlantic Highlands, NJ: Humanities Press, 1996.

Cohen, I. Bernard. "Maxwell's Poetry." *Scientific American* 186 (March 1952): 62–63.

———. *Revolution in Science.* Cambridge, MA: Harvard University Press, 1985.

Cooper, Peter. "Humphry Davy–A Penzance Prodigy." *The Pharmaceutical Journal* 265 (December 23–30, 2000): 920–921.

Crosse, Cornelia. "Science and Society in the Fifties." *Temple Bar* 93 (1891): 33–51.

Davy, Humphry. *The Collected Works of Sir Humphry Davy.* Edited by John Davy. London: Smith, Elder, 1839–1840.

———. *Fragmentary Remains, Literary and Scientific, of Sir Humphry Davy.* Edited by John Davy. London: John Churchill, 1858.

Davy, John. *Memoirs of the Life of Sir Humphry Davy.* London: Smith, Elder, 1836.

Day, Peter. *The Philosopher's Tree: Michael Faraday's Life and Work in His Own Words.* Philadelphia: Institute of Physics Press, 1999.

Devons, Samuel. "The Search for Electromagnetic Induction." *The Physics Teacher* (December 1978): 625–631.

Dibner, Bern. *Alessandro Volta and the Electric Battery.* New York: Franklin Watts, Inc., 1964. (http://dibinst.mit.edu/BURNDY/OnlinePubs/Volta/index.html)

Dyson, Freeman. "Field Theory." *Scientific American* 188 (April 1953): 57–64.

Einstein, Albert, and Leopold Infeld. *The Evolution of Physics: From Early Concepts to Relativity and Quanta.* New York: Simon & Schuster, 1966.

Everitt, C. W. F. *James Clerk Maxwell: Physicist and Natural Philosopher.* New York: Charles Scribner's Sons, 1975.

Faraday, Memoir of His Life. British Museum Add. MSS. 40419, f. 81, 1835. (Attributed to Faraday, but probably not by him.)

"Faraday." *Chemical News* 16 (1867): 110–111.

Faraday, Michael. "Historical Sketch of Electromagnetism." *Annals of Philosophy* 2 (1821): 195–200, 274–290; 3 (1822): 107–121.

——. "Experimental Researches in Electricity." *Philosophical Transactions of the Royal Society of London* 122 (1832): 125–162. [Discovery of electromagnetic induction, based on announcement of November 24, 1831]

——. *The Faraday Lectures.* London: Chemical Society of London, 1928.

——. *Faraday's Diary, Being the Various Philosophical Notes of Experimental Investigation.* Edited by Thomas Martin. London: Bell and Sons, 1932–1936. [**Diary**]

——. *The Chemical History of a Candle.* New York: Viking Press, 1960. (Reprint of original 1861 edition, London: Griffin, Bohn & Co.)

——. *Experimental Researches in Electricity.* New York: Dover Publications, 1965. [**ERE**]

——. *The Selected Correspondence of Michael Faraday.* Edited by L. Pearce Williams. Cambridge, UK: Cambridge University Press, 1971.

——. *Advice to Lecturers: An Anthology Taken From the Writings of Michael Faraday and Lawrence Bragg.* Edited by George Porter and James Friday. London: Mansell, 1974.

——. *Experimental Researches in Chemistry and Physics.* New York: Taylor and Francis, 1991.

——. *The Forces of Matter.* Buffalo, NY: Prometheus Books, 1993. (Reprint of original 1860 edition, London/Glasgow: Richard Griffin.

Faraday's London. London: Royal Institution of Great Britain, 2002.

Feynman, Richard P., Robert B. Leighton, and Matthew Sands. *The Feynman Lectures on Physics.* Reading, MA: Addison-Wesley Publishing Co., 1964

Fisher, Howard J. "The Great Electrical Philosopher." *The College* 31 (July 1979): 1–13.

——. *Faraday's Experimental Researches in Electricity: Guide to a First Reading.* Santa Fe, NM: Green Lion Press, 2001.

Foote, George A. "The Place of Science in the British Reform Movement 1830–50." *Isis* 42 (1951): 192–208.

——. "Sir Humphry Davy and His Audience at the Royal Institution." *Isis* 43 (1952): 6–12.

——. "Science and Its Function in Early Nineteenth Century England." *Osiris* 11 (1954): 438–454.

Forgan, Sophie, ed. *Science and the Sons of Genius: Studies on Humphry Davy.* London: Science Reviews, Ltd., 1980.

Friedel, Robert W. *Lines and Waves: An Exhibit by the IEEE History Center.* New York: Institute of Electrical and Electronics Engineers.

www.ieee.org/organizations/history_center/general_info/lines_men
u.html.1981.

Fullmer, June Z. "Davy's Biographers: Notes on Scientific Biography."
Science 155 (1967): 285–291.

——. *Young Humphry Davy: The Making of an Experimental Chemist.*
Philadelphia: American Philosophical Society, 2000.

Gill, A.J. "Faraday and Photography." *Proceedings of the Royal Institution of
Great Britain* 42 (1967): 54–67.

Gingras, Yves. "What Did Mathematics Do to Physics?" *History of Science* 39 (2001): 383–416.

Gladstone, J.H. *Michael Faraday.* London: Macmillan and Co., 1873.

Glashow, Sheldon L. *From Alchemy to Quarks.* Pacific Grove, CA:
Brook/Cole Publishing Co., 1994.

Goldman, Martin. *The Demon in the Aether: The Story of James Clerk
Maxwell.* Edinburgh: Paul Harris Publishing, 1983.

Golinski, Jan. "Humphry Davy's Sexual Chemistry." *Configurations* 7
(1999): 15–41.

Gooding, David. "Conceptual and Experimental bases of Faraday's
Denial of Electrostatic Action at a Distance." *Studies in History and
Philosophy of Science* 9 (1978): 117–149.

——. "Experiment and Concept Formation in Electromagnetic Science
and Technology in England in the 1820s." *History and Technology* 2
(1985): 151–176.

——. "Scientific Discovery as Creative Exploration: Faraday's Experiments." *Creativity Research Journal* 9 (1996): 189–205.

Gooding, David, and Frank A.J.L. James, eds. *Faraday Rediscovered: Essays on the Life and Work of Michael Faraday, 1791–1867.* Basingstoke,
Hants, England: Macmillan Press, 1989.

Gorman, Mel. "Faraday on Lightning Rods." *Isis* 58 (1967): 96–98.

Hamilton, James. *Faraday: The Life.* London: HarperCollins Publishers,
2002.

Hamilton, James, ed. *Fields of Influence: Conjunctions of Artists and
Scientists, 1815–60.* Birmingham, UK: University of Birmingham,
2001.

Hare, E. "Michael Faraday's Loss of Memory." *Proceedings of the Royal
Institution* 49 (1976): 33–52.

Harman, P.M. "Maxwell and Faraday." *European Journal of Physics* 14
(1993): 148–154.

——. *Energy, Force, and Matter: The Conceptual Development of Nineteenth-Century Physics.* Cambridge, UK: Cambridge University Press, 1997.

Hartley, Harold. *Humphry Davy.* London: Thomas Nelson and Sons, 1966.

Hays, J. N. "The London Lecturing Empire, 1800–50." In *Metropolis and Province, Science in British Culture, 1780–1850,* edited by Ian Inkster and Jack Morrell. Philadelphia: University of Pennsylvania Press, 1983, pp. 91–119.

Helmholtz, Hermann von. "On Faraday." *Nature* 3 (1870): 51–52.

Henry, Joseph. *The Papers of Joseph Henry.* Edited by Nathan Reingold. Washington, DC: Smithsonian Institution Press, 1972–1979.

Hobson, Art. "Teaching E = mc^2: Mass Without Mass." *The Physics Teacher* 43 (February 2005): 80–82.

Holton, Gerald. "The Two Maps: Oersted Medal Response at the Joint American Physical Society American Association of Physics Teachers Meeting, Chicago, 22 January 1980." American Journal of Physics 48 (1980): 1014–1019.

Howe, Michael J. A. *Genius Explained.* Cambridge, UK: Cambridge University Press, 2001.

James, Frank A. J. L. "The Physical Interpretation of the Wave Theory of Light." *British Journal for the History of Science* 17 (1984): 47–60.

——. "Michael Faraday, the City Philosophical Society and the Society of the Arts." *Royal Society Arts Journal* 140 (1992a): 192–199.

——. "The Tales of Benjamin Abbott: A Source for the Early Life of Michael Faraday." *British Journal for the History of Science* 25 (1992b): 229–240.

——. "Faraday, Maxwell, and Field Theory." In *Semaphores to Short Waves.* London: Royal Society of the Arts (1998): 71–84.

——. *Guides to the Royal Institution of Great Britain: 1. History.* London: Royal Institution of Great Britain, 2000.

James, Frank A. J. L., ed. *The Correspondence of Michael Faraday,* vols. 1–4. London: Institution of Electrical Engineers, 1991–1996.

Jeffreys, Alan. *Michael Faraday, A List of His Lectures and Published Writings.* London: Chapman and Hall, Ltd., 1960.

Jones, Henry Bence. *The Life and Letters of Faraday,* vols. I–II. Philadelphia: Lippincott, 1870. [**BJI, BJII**]

Knight, David. *Humphry Davy: Science & Power.* Cambridge, MA: Blackwell Publishers, 1992.

———. *The Collected Works of Sir Humphry Davy.* Bristol, UK: Thoemmes Continuum. www.thoemmes.com/science/davy_intro.htm. 2001.

Kondo, Herbert. "Michael Faraday." *Scientific American* 189 (October 1953): 91–96.

Larmor, Joseph. *Origins of Clerk Maxwell's Electrical Ideas as Described in Familiar Letters to William Thomson.* Cambridge, UK: Cambridge University Press, 1937.

———. "Faraday on Electromagnetic Propagation." *Nature* 141 (1938): 36–37.

Lindee, M. Susan. "The American Career of Jane Marcet's *Conversations on Chemistry,* 1806–1853." *Isis* 82 (1991): 8–23.

MacDonald, D. K. C. *Faraday, Maxwell, and Kelvin.* Garden City, NY: Anchor Books, 1964.

Marcet, Jane Haldimand. *Conversations on Chemistry.* 9th American edition. Hartford, CT: Oliver D. Cooke and Sons, 1824.

Martin, Thomas. *Faraday's Discovery of Electro-Magnetic Induction.* London: Arnold, 1949.

Maxwell, James Clerk. "On Faraday's Lines of Force." *Transactions of the Cambridge Philosophical Society* 10 (1856): 27–83.

———. *A Treatise on Electricity and Magnetism.* London: Geoffrey Cumberledge, 1904.

———. *The Scientific Letters and Papers of James Clerk Maxwell.* Edited by P. M. Harman. Cambridge, UK: Cambridge University Press, 1990–2002.

Mertens, Joost. "Shocks and Sparks: The Voltaic Pile as a Demonstration Device." *Isis* 89 (1998): 300–311.

Miller, David Philip. "Between Hostile Camps: Sir Humphry Davy's Presidency of the Royal Society, 1820–1827." *British Journal for the History of Science* 16 (1983): 1–47.

Morus, Iwan Rhys. "The Sociology of Sparks: An Episode in the History and Meaning of Electricity." *Social Studies of Science* 18 (1988): 387–417.

———. "Currents from the Underworld: Electricity and the Technology of Display in Early Victorian England." *Isis* 84 (1993): 50–69.

———. *Michael Faraday and the Electrical Century.* Cambridge, UK: Icon Books, 2004.

Nersessian, Nancy J. "Reasoning from Imagery and Analogy in

Scientific Concept Formation." *Proceedings of the Biennial Meeting of the Philosophy of Science Association* 1 (1988): 41–47.

Newman, James R. "James Clerk Maxwell." *Scientific American* 192 (June 1955): 58–71.

Nielsen, J. Rud. "Hans Christian Oersted–Scientist, Humanist and Teacher." *Physics History from AAPT Journals.* Edited by Melba Newell Phillips. College Park, MD: American Association of Physics Teachers, 1985, pp. 23–35.

Nye, Mary Jo. *Before Big Science: The Pursuit of Modern Chemistry and Physics 1800–1940.* Cambridge, MA: Harvard University Press, 1996.

Paris, John Ayrton. *The Life of Sir Humphry Davy.* London: Colburn and Bentley, 1831.

Pauling, Linus. *General Chemistry.* San Francisco: W. H. Freeman and Co., 1953.

Purrington, Robert D. *Physics in the Nineteenth Century.* New Brunswick, NJ: Rutgers University Press, 1997.

Ross, Sydney. "Scientist: The Story of a Word." *Annals of Science* 18 (1962): 65–85.

——. "John Herschel on Faraday and on Science." *Notes and Records of the Royal Society of London* 33 (1978): 77–82.

Sagan, Carl. *The Demon-Haunted World.* New York: Random House, 1995.

Seeger, Raymond J. "Michael Faraday and the Art of Lecturing." *Physics Today* 21 (1968): 30–38.

Shamos, Morris. *Great Experiments in Physics.* New York: Holt, Rinehart and Winston, 1959.

Simpson, Thomas K. "Faraday's Thought on Electromagnetism." *The College* 22 (July 1970): 6–16.

——. *Maxwell on the Electromagnetic Field.* New Brunswick, NJ: Rutgers University Press, 2001.

Thomas, John M. *Michael Faraday and the Royal Institution: The Genius of Man and Place.* Bristol: Adam Hilger, 1991.

Thompson, Silvanus P. *Michael Faraday, His Life and Work.* London: Cassell and Co., 1898.

Tolstoy, Ivan. *James Clerk Maxwell: A Biography.* Edinburgh: Canongate Publishing, 1981.

Treneer, Anne. *The Mercurial Chemist: A Life of Sir Humphry Davy.* London: Methuen & Co. Ltd., 1963.

Tricker, R. A. R. *The Contributions of Faraday and Maxwell to Electrical Science.* New York: Pergamon Press, 1966.

Tunbridge, Paul A. "Faraday's Genevese Friends." *Notes and Records of the Royal Society of London* 27 (1973): 263–298.

Tyndall, John. "On the Existence of a Magnetic Medium in Space." *Philosophical Magazine* 9 (1855): 205–209.

——. *Faraday as a Discoverer.* New York: Thomas Y. Crowell, 1961. (Reprint of original 1868 edition, London: Longmans, Green & Co.)

——. *Fragments of Science*, vol. I. London: ElecBook, 2001. (Reprint of original 1879 edition, London: Longmans, Green & Co.)

Tyrell, H. J. V. *Guides to the Royal Institution of Great Britain: 2. The Site and the Buildings.* London: Royal Institution of Great Britain, 2001.

Watson, E. C. "On the Relations between Light and Electricity." *American Journal of Physics* 25 (September 1957): 335–343.

Watts, Isaac. *The Improvement of the Mind; or, a Supplement to the Art of Logic.* London: William Baynes, 1809.

Williams, L. Pearce. "Humphry Davy." *Scientific American* 202 (1960a): 106–116.

——. "Michael Faraday's Education in Science." *Isis* 51 (1960b): 515–530.

——. *Michael Faraday.* New York: Basic Books, 1965.

——. "Michael Faraday: A Biography." *British Journal for the Philosophy of Science* 18 (1967a): 230–240.

——. "Michael Faraday and the Ether: A Study in Heresy." *Proceedings of the Royal Institution of Great Britain* 41 (1967b): 666–680.

——. "Should Philosophers Be Allowed to Write History?" *British Journal for the Philosophy of Science* 26 (1975): 241–253.

Wise, M. Norton. "The Mutual Embrace of Electricity and Magnetism." *Science* 203 (March 30, 1979): 1310–1318.

Yarrow, A. F. "An Incident in Connection with Faraday's Life." *Proceedings of the Royal Institution* 25 (1926–28): 480.

ACKNOWLEDGMENTS

I acknowledge the generous assistance of George Gibson, Jackie Johnson, and Michele Amundsen at Walker Books; my agent Sally Brady; Faraday scholars L. Pearce Williams and Frank A.J.L. James, who freely provided their expertise; Wolf Kern, who read a draft version of the manuscript; my colleagues at UMass Dartmouth, for their patience and support; Harvard College Observatory, for the courtesy appointment that allows me access to Harvard's vast library resources; and my wife, Sasha, who was always there when I needed the reader's opinion.

ART CREDITS

INDEX

electricity (*continued*)
of, 205; Faraday's contributions to
understanding, 154; Faraday's
work on, 8–9, 18–21, 128, 163,
185–86, 192–93, 205; fluid
theories of, 8, 9, 11–12, 132, 133,
139, 140, 144–45; frictional process of, 113; gravity converted
into, 128, 205; induction of,
113–24, *114, 116*, 126–31, 137,
140–41, 170, 174, 188; as inherent
power of matter, 140; Maxwell's
work on, 188, 190; nature of,
132–34; as nonmaterial
phenomenon, 136; and science in
early nineteenth century, 4; and
solid-liquid matter, 133–34; and
spiritualism/pseudoscience, 196;
static, 12, 113, 132, 139; Tatum's
lectures about, 10–11, 12, 15, 16;
and time of magnetic and electric
movement through space, 185–86;
Tytler's essay about, 8, 11, 12; as
universal force, 160; voltaic, 132,
136–37, 170. *See also*
electrochemistry; electrolysis;
electromagnetism; electrostatics;
specific person
electrochemistry: Davy's interests in,
32–35; emergence of field of, 12;
Faraday's challenge in, 145;
Faraday's experiments involving,
18–21, 93; and hulls of warships
experiments, 93; as promising path
of research, 18; vocabulary
concerning, 135
electrolysis: Faraday's studies about,
134–36
electromagnetism: and currents of
electricity, 129; Einstein's views
about Faraday's theory of, 212;
Faraday's contributions to
understanding, 154; Faraday's
lectures about, 120, 124, 125, 126;
Faraday's work on, 73–85, *81, 82,*

87, 112–24, 126, 154–60; Faraday's
writings about, 125, 126, 130;
French reports about Faraday's
work with, 126; induction of,
113–24, 126–31, 137, 174; and
light, 154–60, 165–73, 188–90;
and lines of force, 127–30, *128,*
166; and mathematics, 165, 190;
Maxwell's work on, 181, 182–83,
188–90, 191–93; and nature of
"electricities," 132–34; and
Wheatstone's stopwatch, 165. *See
also* electrochemistry; electrolysis
electrostatics: Faraday's studies
about, 138–49, *143*
elements, chemical: Davy's early
interests in study of, 24–25;
Davy's isolation of, 18, 21, 35. *See
also specific element*
Eliot, T. S., 211
Emerson, Ralph Waldo, 150
ether: luminiferous, 166, 169
Experimental Researches in Electricity
(Faraday), 125, 132, 134, 136, 144,
146–47, 153, 154, 181, 182, 192–93
experiments: as basis of Faraday's
procedural philosophy, 21; dangers
of, 26, 36, 38, 85; Davy's
inhalation, 25–28; of Faraday in
1850s and 1860s, 205; and
Faraday's creative process in late
stages of his career, 205; Faraday's
early, 5–6, 8–9, 18–21; Faraday's
last, 209; Faraday's views about
importance of, 205, 212;
Maxwell's views about, 181, 182.
*See also specific scientist or type of
experiment*

facts: Faraday's views about, 13–14,
17
farad (measure of charge-storing
capacity), 136
"Faraday cage," 141–42
Faraday, James (father), 3, 4